A149
LANDMARKS

Edward Couzens-Lake

EDWARD COUZENS-LAKE

AMBERLEY

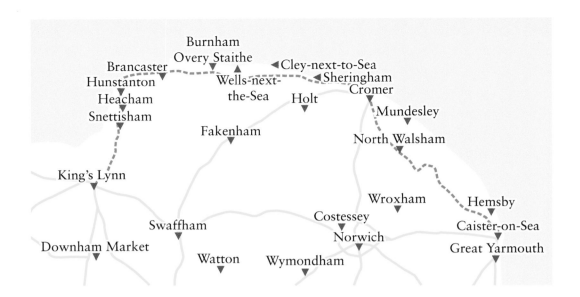

To Dad
1938–2015

First published 2017

Amberley Publishing
The Hill, Stroud
Gloucestershire, GL5 4EP

www.amberley-books.com

Copyright © Edward Couzens-Lake, 2017

The right of Edward Couzens-Lake to be identified
as the Author of this work has been asserted in
accordance with the Copyrights, Designs and
Patents Act 1988.

ISBN 978 1 4456 6162 9 (print)
ISBN 978 1 4456 6163 6 (ebook)

British Library Cataloguing in Publication Data.
A catalogue record for this book is available from
the British Library.

Typeset in 10pt on 13pt Celeste.
Typesetting by Amberley Publishing.
Printed in the UK.

Contents

Introduction

The A149 has always played a significant role in my life.

I would have first travelled along its winding path as, cradled in my parents' arms, I made my way to my childhood home in Brancaster for the first time. It's the road that would have been part of my daily journey to the two Norfolk schools and one college I attended as a child, as well as the one that took me to King's Lynn station and the start of a new life in London.

There were trips aplenty to see the doctor, to go to church, to visit the many family members scattered along its route, for nights out at one of a handful of glorious pubs, or to visit seaside towns and villages with the intent to explore, either on my own or with friends, as well as for a myriad of holiday jobs. All involved a journey, long or short, along the A149 – the road my bride walked along to the village church on our wedding day. That same road and church I quietly walked for the funeral of my much-loved and still-missed father some years later; he who would, endlessly and without complaint, give me lifts to and from all the places I visited on that stretch of highway we all call 'The Coast Road'.

In an ever-changing life and a world that expects us all to grow up and move on rather more swiftly than we might like, it is and remains an absolute constant; a letter followed by three numbers that always reminds me that I am nearly home when I first see it appear on one of those garish green roadway signs of which the Highways Agency are so fond.

Me and countless thousands of other people. One road, but so many journeys to be had and so many more yet to be taken. Our highway, our lives. This book is my little homage to the A149. It features seventy-five landmarks along the way, either readily seen or just behind a Norfolk corner as you travel its 84 miles in an easterly direction from King's Lynn towards Great Yarmouth. Some will be familiar, others will not – while one no longer exists! There will also be some, I suspect, of which you question the inclusion, while others you will deem to have been inexcusably omitted. There will also, I have no doubt whatsoever, be facts or figures that you may question or else feel the need to verify. The source material I have worked from is rich, varied and wonderful in content, tone and information – so much so, that some of it contradicts what is written somewhere else! Rest assured that any errors or mistakes made in the text are unintentional and will, once advised, be made good for subsequent reprints of this book.

It is important to realise that this is a book to browse, to peruse in the hope that a fleeting mention, reference or photograph might encourage you to find out more about the place I have written about. The descriptions given should not be considered as definitive historical accounts of the places named but as personal observations and musings mixed in with a few facts, some of which will almost certainly be apocryphal – and delightfully so.

Historian I most certainly am not. What I am is a man constantly in love with all things Norfolk.

Some of the places mentioned may not even be on the A149 itself but will require a detour from its sinuous route. Rest assured that, if that is the case, then the journey off track will be a short but worthwhile one to make.

Enjoy your journey and some of these destinations you may stop at to take a closer look.

Along the old Coast Road.

Edward Couzens-Lake

Foreword by Nick Conrad, BBC Radio Norfolk

I love maps. Norfolk roads snake across our county like varicose veins across skin. Like veins, some arterial routes are more prominent than others. The A149 is one such important road. This super-highway happens to be my favourite piece of prepared tarmac in the whole of the UK. Why? I can chart my whole life story along it.

If I want to reminisce, I only need a few hours and a tank of petrol. Trundling along you could be transported back in time. I imagine little has changed in some villages since the 1950s. In fact, so relaxing and therapeutic is the A149 for me, it's like a free therapy session. In this foreword, can I entice you to join me on a gentle little journey?

Let's start in the Burnhams. I fondly remember many happy memories of picnics on the beach. This 'trendy' part of Norfolk has changed quite a bit recently but it still holds a unique charm. West Norfolk is not rich! But this corner of our county is called Chelsea-on-Sea by some. Welcome to the land of 4x4s, boat owners and children called

Tarquin or Tallulah! You're just as likely to get stuck behind a Wayfarer (a type of sailing boat) as you are a caravan or tractor. In fact the coast road can be a 'seasonal slalom', avoiding a collision with the trappings of other people's wealth. If it isn't clipping an outboard motor on the latest speedboat, it's a horse box or brand new Porsche that warrants careful manoeuvre. The beach here is worth the long walk. At low tide the sea seeps out as far as the eye will muster vision. As a youngster, I despaired as the water disappeared; I imagined Paul Daniels would jump out as if he'd performed a trick to create the distance exposed by low tide. Yet the absence of the sea still leaves an inviting creek, which on a hot day wills you to jump in and enjoy its icy refreshment.

Wells is a warm, friendly little town and always worth a visit. I accept that you'll have to battle to find a parking space but, when you have, what a treat. The Quayside is a lovely place to relax with a hot steaming portion of fish and chips.

Now to Norfolk's gems. Stiffkey is famed more for its eccentric former vicar than for its intoxicating sleepiness. As a boy I was intrigued by the wonderful story of the Reverend Harold Davidson. I have visited his grave on countless occasions, drawn by his unbelievable story. It's a terrific tale! Are you sitting comfortably? Then I'll begin ... Davidson, the Rector of Stiffkey, was perhaps better known as the Prostitutes' Padre. He was thrown out of the Church for 'immoral behaviour' in the 1930s. But was he the victim of a terrible miscarriage of justice? The Rector led a double life. As well as overseeing his congregation in Stiffkey, he commuted to Soho where he ministered to a rather different flock. These were down-and-outs and prostitutes, whom he had helped for years.

The kind-hearted commuting Rector ran into trouble when he was late arriving back from soul saving in Soho for the annual Remembrance Day service. The Bishop of Norwich was (allegedly) incandescent with rage! Doubting the reason for his Soho visits, the Bishop's lawyer had a private detective agency crawling around country lanes and peering into Soho café windows for signs of Harold and girls of easy virtue. It was a fix, and evidence was 'constructed' against Harold – he faced five counts of immoral behaviour. Life became much worse for Harold following his banishment from the Church. In an attempt to raise money for an appeal, he exhibited himself in a barrel at Blackpool. He then joined a fair to preach from the lion's den. Sadly he trod on the lion's tail, leading to a swift turn of events that ultimately saw him meet his end!

Seventy years after his funeral, his epitaph reads, 'He was loved by the villagers who recognised his humanity and forgave him his transgressions. Rest in peace.'

Heading on towards Morston, where my father taught us how to sail, brings back so many memories. Salt spray in our faces, seals dancing around the boat and soggy Wagon Wheels. As you drive along the A149 past Cley next the Sea, open up your nose! Yes, breathe in deeply.

One smell always reminds me of home. As you can imagine, a lot of my time is spent in London. Us broadcasters tend to spend a significant amount of time in our beloved capital. Encased in our air-conditioned studios polluted by the smell of potpourri air freshener. So, whenever I come home, imagine my delight when my nose begins to tingle, as I approach that familiar smell reminding me of home.

So from where does this smell emanate? Usually around one mile inland you get the first waft. Carried on the breeze, the undervalued, gorgeous and quite individual scent of the North Sea! This is the hard bit. Avoiding clichés like 'salty' or 'fishy', how do you describe the smell of seawater? It's a sweet, fresh, yet heavy smell that lingers and wafts. Its mysterious presence heightened by its ability to drift over the fields. It stalks your nose, delighting and offending your senses in equal measure.

But then something splendid happens. When you're elevated above the cliffs where this curious aroma originates – it's gone. Magical. Its intoxicating, almost delicious pungency is erased by the stiffest of refreshing breezes. It delights without fail.

I feel like a fish out of water when I'm away from the Norfolk coast. I know I'm starting to sound like a broken record! But I make no apology for being totally enamoured by our gorgeous coastline. I function better when whipped by gusts of salty wind or suffering from the inconvenience of sand in my shoes. Maybe I'm part merman?

But why do we love odd smells? Familiarity. The scientific explanation is deeply embedded in our psychology. That said, no explanation has thus far helped elucidate my appreciation for the stench of manure! Maybe I really am an old country bumpkin!

Sheringham is where I grew up in one of the most delightful communities in the whole county. Cromer with its iconic pier, then onto Yarmouth with all its temptation. I could go on ... but that would defeat the purpose of this book.

Edward is an incredibly talented writer. I'm so pleased he's chosen this particular road to explore. As you can see from my passionate foreword I'm jealous I didn't write this book before him! I'm hugely looking forward to reading about his adventures.

Nick Conrad

❶ South Gates, King's Lynn
Imposing gateway to a historic town.

This impressive gateway to the old southern entry point to King's Lynn is a Grade I listed structure that was originally constructed in the fourteenth century, although much of it was rebuilt in the 1450s – work that was ultimately responsible for much of the building that can be seen today.

The impressive frontage, which is constructed of masonry made of large square-cut stones, is a sixteenth-century facade that protects the original brick and mortar construction, now believed to be one of the oldest surviving buildings of this type in the country. Later modifications were added to the original structure in the nineteenth century; these were the two smaller arches built to the side of the main one. Their intended use was for the lower classes to enter or exit the town, whilst the main arch was exclusively intended for the use of horse-drawn carriages. This ensured that their frequently important occupants would enjoy being noticed, such was the grand and high profile nature of such an entrance to the town.

The South Gates are situated at the top of the road (A10) that leads away from the town and, ultimately, down into London, which means they are near one of the busiest traffic intersections in the area: the Hardwick Roundabout, a scene of fear and foreboding for many a learner motorist from across the region. While the author acknowledges that the South Gates in King's Lynn are situated on the A148, such is their historical significance, together with the visual impact and interest they have for the traveller, they are included in

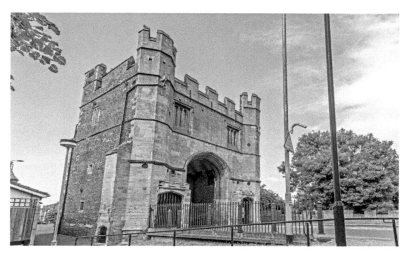

(Borough Council of King's Lynn & West Norfolk)

this book as the 'gateway' to the A149 as it commences its route across the county towards Great Yarmouth.

Despite the constant stream of traffic that flows in and out of the town via the South Gates every day, accidents in or near the gates themselves are infrequent as the size of the main arch comfortably allows easy passage for even the most formidable articulated lorry.

❷ Queen Elizabeth Hospital, King's Lynn
King's Lynn District General Hospital (NHS).

Not a site or destination that most would opt to visit by choice, the Queen Elizabeth Hospital does nonetheless more than earn its place in this collection for the important role it plays in regard to the day-to-day health of not only the residents of West Norfolk but also of those in South Lincolnshire and North East Cambridgeshire, an area that has a population of approximately 250,000 people. The current monarch has had good reason to make use of the hospital that bears her name, not least in 2003 when, while staying at the royal residence at Sandringham (see p. 6), HM the Queen was taken to the hospital after having experienced some discomfort in one of her knees. She underwent a scan before being transferred to the King Edward VII Hospital in London where, in the facility named after her great-grandfather, an operation to remove a torn cartilage was carried out.

The main hospital building on the site is comprised of two storeys; it was completed in 1980 and named after the late Queen Mother, and not, as is regularly and inaccurately claimed, the present Queen.

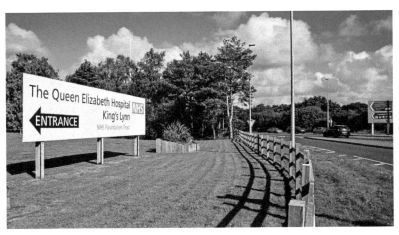

(Nigel Nudds)

The original site has since been expanded to include several newer buildings. These include the Fermoy Unit, an adult mental health facility that also opened in 1980, as well as the Roxburgh Children's Day Centre. Although the site is primarily NHS managed (by the Queen Elizabeth Hospital King's Lynn NHS Foundation Trust) and funded, it also contains a private BMI site known as the Sandringham Hospital.

The hospital's previous site at London Road in King's Lynn was the birthplace of the author.

③ Ruins of Church of St James at Bawsey, near King's Lynn
Site of long-abandoned Norman church.

The sight of this sad and desolate ruin, adrift in the middle of an otherwise featureless field, has never failed to capture my imagination, so much so that I now look out for it on every occasion that I am in the area and wonder, perhaps with too much romantic licence, whatever might have happened to consign that lonely church to such an ignominious state.

As it turns out, the ancient church of St James would have been the central feature of the now long-gone hamlet of Bawsey St James up until the sixteenth century when, in a gesture not completely unique at the time, the then landowner of the settlement decided that its value to his estate was barely worth bothering about anymore. Thus, with scant regard to the people that lived there, he turned the whole hamlet over to farmland, razing to the ground any and all of

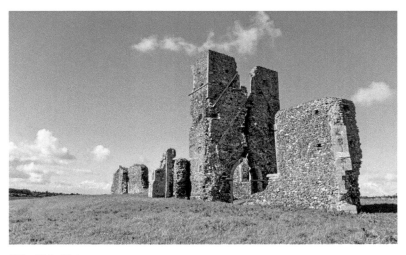

(Nigel Nudds)

the buildings in the hamlet apart from the church, whose ruins stand out to this day in the field that was once valued above human lives.

Such is the manner in which the remains of the church stand out against the big Norfolk skies and the otherwise rather featureless stretch of the A149 in which they stand, it is likely that they are noted and commented upon by countless thousands of travellers every year. Yet very few take the time and trouble to visit its isolated hilltop location. Among those who did were archaeologists from Channel 4's popular *Time Team* programme who, when exploring the surrounding area in 1999, discovered, among other items, the skeleton of a one-time inhabitant whose skull had clearly received a fatal blow from an anonymous swordsman.

Perhaps the residents of Bawsey did not take leave of their village as meekly as the landowner might have expected?

❹ Castle Rising Castle, Castle Rising
Twelfth-century medieval fortification once owned by Queen Isabella of France.

Although the pretty village of Castle Rising and its castle lie a little under a mile north-west of the A149 as it heads towards the royal home and grounds of Sandringham, it is a site that demands attention and is well worth a slight deviation in order to pay a visit – one that can, quite easily, take up half a day without the curious explorer even being aware that so much time has passed.

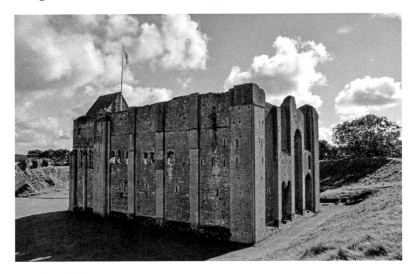

(Nigel Nudds)

The keep (a fortified tower built within the confines of the outer castle walls) at Castle Rising is one of the largest and best preserved in England. It is now surrounded by twenty acres of formidable earthworks, designed to both dissuade and, if that didn't work, ultimately repel dissenters and the attacks of enemy soldiers.

These earthworks further protect three inner courtyards that, in total, cover an area of around twelve acres. The keep itself, situated in the innermost courtyard, is said to have been based upon the great keep of Norwich Castle with the one at Rising, despite its more modest location and reputation, reckoned by some historians to surpass Norwich as one of the finest of all Norman keeps. Quite an accomplishment for a castle in a quiet and otherwise very understated Norfolk village.

As with many castles of the time, it was built on elevated ground not only to provide a natural defensive advantage over the surrounding countryside but also to emphasise the power and influence of its residential lord over that countryside and its population – an exercise in ego built from solid and unyielding stone. The castle and surrounding lands are currently managed by English Heritage.

5 Babingley

Hamlet and abandoned village, now part of the civil parish of Sandringham.

The modern-day hamlet of Babingley is no more than a small group of houses scattered along the A149, around seven miles NE from King's Lynn. Our interest here, however, is in the abandoned village that lies in fields to the west of the road, one marked by the ruins of St Felix's parish church.

Babingley is claimed to be the site where St Felix of Burgundy, the Apostle to the East Angles, landed in England in around AD 615, his arrival leading to an invitation from the area's ruling family to preach Christianity throughout the region. Popular legend supposes that Felix's arrival at Babingley was more through accident than design, having come about as the result of his being shipwrecked on the river that shares the hamlet's name. Felix, saved by a well-meaning colony of beavers, was so grateful for their mercies that he immediately consecrated one of the beavers as a bishop. This act is commemorated on the village sign, which shows a grateful St Felix handing a bishop's mitre to the beaver.

The ruins of the fourteenth-century St Felix's Parish Church are said to be on the site of the first Christian church to be built in

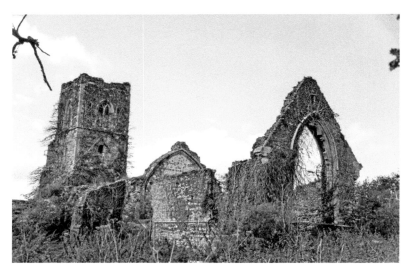

(Nigel Nudds)

Norfolk. It was regularly used for worship for nearly 500 years before falling into its current state of disrepair. The site is now part of the nearby Sandringham estate (see the following entry) and can only be accessed through the express permission of the landowner.

The curious explorer might also want to look for Butler's Cross. This is the base and broken shaft of a medieval stone boundary cross that is sited on a traffic island at the junction where the B1439 to West Newton meets the A149. Its name is derived from that of the de Botler family, who held the manor of West Hall in Babingley from the mid-thirteenth century.

⑥ The Sandringham Estate
The Norfolk country retreat of HM the Queen and HRH the Duke of Edinburgh. Includes the house, country park, museum and visitor centre.

As the A149 heads towards the coast, it meanders its way through the woodland, plantations and open countryside of the Sandringham Estate. This is an area of approximately 20,000 acres of land near the village of the same name, of which the focal point is Sandringham House, a private residence owned by HM the Queen.

The estate on both sides of the road is the overall responsibility of HRH the Duke of Edinburgh, who took on the management of the site in 1952. Despite his reputation for being very much a 'hunting, shooting and fishing' man, the Duke's overriding priority for the Sandringham

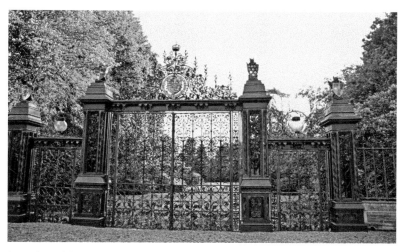

(Nigel Nudds)

Estate has been that of conservation, ensuring that it can be enjoyed by visitors for countless generations to come. This means a commitment to planting several thousand sapling trees on an annual basis, as well as new hedgerows and the creation of wetland areas.

The intriguingly named Cats Bottom, situated within the estate margins, is an area of conifer woodland that also consists of a few cottages built, it is thought, for workers at the estate's gravel pits. It has, perhaps with the intention of soothing the sensitivities of those who might be easily offended, found itself called 'Catchbottom' in recent years, but this would seem to be nothing more than latter-day censorship as the location is always marked as Cats Bottom on old maps.

The log cabin-type building that can be seen on the inland side of the A149 as one travels through the estate and towards Dersingham is owned by the Babingley Social Club. The club is over 100 years old, having been opened by King George V and Queen Mary in 1913. They were accompanied by a star-studded supporting cast of fellow royals on the day with Queen Alexandra, Princess Victoria, and the King and Queen of Norway.

7 Wolferton Station
Destination of royal visitors to Sandringham House.

As with some other locations in this book, Wolferton station is situated just off the A149, in this instance just over a mile west of the road as it wends its way to the coast via Dersingham and Ingoldisthorpe.

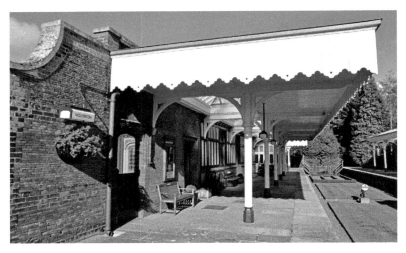

(Nigel Nudds)

The planned Wolferton station, intended to be open and ready for regular use in 1862 as part of the Lynn & Hunstanton Railway's route between the two towns, suddenly found itself located in a prime site following the purchase of the nearby Sandringham Estate as the intended new home of the then Prince of Wales. One can only speculate at the delight that must have swept over the company's directors as they heard the news and the realisation that, far from being, even for that time a small and relatively obscure station in the middle of nowhere, their new project was now going to find itself designated as the 'royal station', a description that has remained with it to the current day, even though the station itself was closed in 1969.

The directors didn't immediately set about transforming their hitherto humble station in order to make it suitable for visiting royals, however, with the station not seeing much in the way of redevelopment until 1898, when some incongruous Tudor-style platform buildings were constructed. These included the newly designated 'royal waiting rooms', which were fitted out with expensive oak panelling, couches and easy chairs. Even the smallest of detail was scrupulously attended to with the already ornate platform lamps being topped with miniature crowns.

The King's Lynn–Hunstanton line, and with it Wolferton station, closed in 1969 with, as a parting gesture, the present-day Queen being offered the royal waiting rooms – for purchase, mind, rather than a gift. HRH declined the offer, agreeing, as she did so, that the station at King's Lynn would now be her and the family's preferred local station. The station was retained as a museum until 2014, when it was sold to a private buyer. Access to the station and surrounding area is therefore dependent on receiving permission to do so from the current owners.

16

Avid viewers of the Netflix television series *The Crown* may have wondered about the location of the spectacular viaduct over which the Royal Train was frequently shown travelling en route to Wolferton station. Rest assured that, while the station still very much exists, the viaduct was nothing more than a figment of the programme-makers' imagination.

⑧ St Mary's Church, Snettisham
'Perhaps the most exciting decorated church in Norfolk' – Nicholas Pevsner.

Norfolk has its fair share of spectacular village churches, with St Mary's at Snettisham being one of the grandest of all. Approaching it from the west and the A149, which bisects the village, its vista is hugely impressive with its spire, at 175 feet, redolent of its mighty sister at Norwich Cathedral.

Even if you don't have time to visit the church itself, do take a few moments to consider the numerous qualities of this fourteenth-century building, its size a representation of how important the village was at the time of its construction. This is a church that so impressed the planners of the Christ Church Cathedral in Fredericton, Canada, that they modelled their own building on St Mary's and refer to Snettisham today as 'the Mother church'.

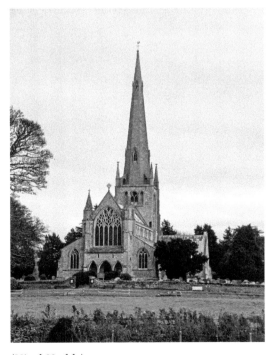

The church was a more-than-tempting target for the commander of one of the German Zeppelin airships that took part in the first ever airborne raid on England in 1915. As Snettisham was not subject to a 'blackout' at the time (hardly surprising, given its relatively remote location and lack of any strategic importance), the scale and obvious visibility of the church led to a bomb being dropped on it – one that, fortunately, missed its target and exploded on a nearby pasture. The force of the explosion was still enough to cause some damage, including to a window on the church's east side. St Mary's in Snettisham can therefore claim to be one of the very first buildings in the world to be bombed from the air.

The spire was often used as a navigation point for sailors navigating the nearby Wash, and was once referred to as 'God's rocket to heaven', such was its prominence on the horizon.

(Nigel Nudds)

⑨ Village Sign – Heacham
Traditional Norfolk village sign that celebrates Pocahontas.

Heacham proudly celebrates its link with the Native American princess Pocahontas by including her image on its village sign, as well as on a plaque inside the village church.

Pocahontas briefly visited the village along with her husband, Sir John Rolfe, in 1616, with Rolfe taking the opportunity to show her his estate at Heacham Hall. They did not, however, settle into married life in Norfolk but departed after their brief stay to Brentford where they lived until her death a year later in Gravesend.

Heacham once had a thriving brick industry, which prospered up until 1862 when the village's railway station was opened. This meant that cheaper (and poorer quality) bricks were more easily obtainable and could be transported in by rail from the ports at King's Lynn and Wells. The arrival of the station, however, was not so much to end Heacham's 'industrial age' as to prepare it for life as part of the thriving holiday market along the Norfolk coast at that time. It duly prospered in its new role and remains popular to this day with visitors, many of whom elect for the quieter charms of its beach rather than the more rumbustious surroundings of Hunstanton's sea front a little further north along the coast.

Some readers may remember Albert, Heacham's friendly village barber who, along with my father, I regularly visited – another reason to take a trip along the A149.

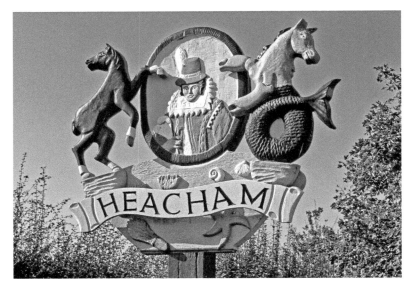

(Mark Oakden at www.tournorfolk.co.uk)

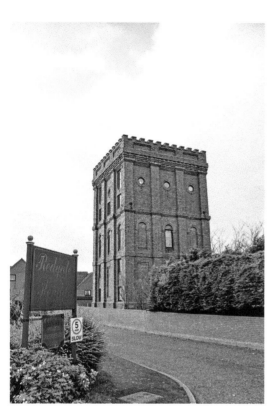

(Nigel Nudds)

⑩ Redgate Water Tower, Hunstanton
Distinctive Edwardian water tower.

A pressing need to provide for both a growing resident population as well as ever increasing numbers of holiday visitors to Hunstanton at the beginning of the twentieth century saw Hunstanton Urban District Council commission the building of this eye-catching water tower in 1912. The council nobly following the lead set by Victorian engineers of making buildings intended to be utilitarian in nature but as attractive and striking in design as their imaginations would allow.

Named after the hill upon which it stands, the Redgate water tower was designed and built to supply water to the adjacent village of Heacham, where an additional five miles of mains pipes were also laid. The tower, which had a peak capacity of 50,000 gallons, was known locally as the Heacham tower with that hilltop location providing a top water level of 160 feet.

The tower was saved the ignominious fate suffered by its equally distinct comrade in Hunstanton itself. That water tower was built right on the edge of the A149 as it skirted the outskirts of the town on the corner of Lincoln Street and Cromer Road but was, to my great sadness at the time, demolished in the early 1980s. Redgate water tower was converted into private apartments in 1984.

⑪ Smithdon High School, Hunstanton
Uncompromising early example of 'brutalist' architecture.

The long and somewhat stark vista of Smithdon High School cannot easily be missed by the everyday motorist as he/she drives northwards along the A149 en route to Old Hunstanton and the freely scattered coastal villages that populate it along Norfolk's north-facing coast, home to many of its far-reaching intake of pupils aged from eleven to eighteen years old. The school was originally constructed almost entirely of glass and steel from a design borne of the architectural talents of Peter and Alison Smithson, seen as the first example of 'brutalist' architecture

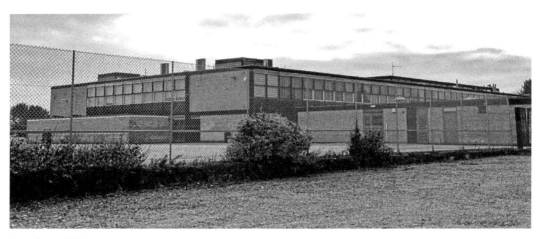

(Nigel Nudds)

at its official opening in 1955. It is now a Grade II listed building that is protected from any substantial additions or alterations without the approval of English Heritage.

Though the design may have been a revolutionary one, the mix of glass and steel did cause a few practical problems, to which, speaking as former pupil at the school, I can attest. Temperatures in the summer months would often be uncomfortably warm. The drought-hit summer of 1976 with its attendant plagues of ladybirds was an additional issue for staff and pupils to contend with while, on the other hand, coats and scarves would often be needed in the winter to make conditions in the classrooms at that time of year even vaguely bearable. Some of these heat-related problems have since been alleviated by the replacement of some of the large glass panes with similarly sized black panels.

The school's library has recently been refurbished to provide twenty-one new internet-connected PCs, while its gymnasium is listed under the Planning (Listed Buildings and Conservation Areas) Act of 1990 as having special architectural or historic interest.

⑫ Old Hunstanton Lighthouse and Ruins of St Edmund's Chapel, Hunstanton
Nineteenth-century clifftop lighthouse and adjacent ruins of ancient chapel.

Hunstanton's iconic lighthouse can easily be seen on the A149 as it heads away and east of the town, before turning eastwards for that famous long stretch of coastal road that extends from Old Hunstanton through to Cromer.

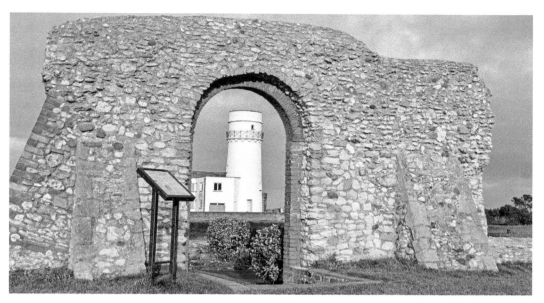

(Nigel Nudds)

The current building dates from 1840, although there has been a lighthouse on the current site since 1665. The first to be constructed there was a primitive but effective structure made from timber with an iron basket atop that generated its light through the burning of coals. This was superseded in 1776 when a new lighthouse had the distinction of generating its light through the installation of the world's first parabolic reflector. The site and building were acquired by Trinity House (the official General Lighthouse Authority for England, Wales, the Channel Islands and Gibraltar) in 1837, and it was they who designed and constructed the present lighthouse, which was in use from 1840 until deactivation in 1921. The former lighthouse is currently utilised as a holiday let.

No picture or visit to the lighthouse can be taken or enjoyed without the casual explorer noting the nearby ruins of what was St Edmund's Chapel. It was built in 1272 in memory of St Edmund, who landed at Hunstanton in 855, becoming King of the East Angles (Anglia) in the process. Edmund later led an army against marauding Viking invaders but was defeated, captured and martyred. He later became the first patron saint of England and is, perhaps, a worthier owner of that epithet than St George.

Excavations carried out on the site in 1913 saw the recovery of many artefacts, including fragments of fifteenth-century window glass and glazed floor tiles from the fifteenth and sixteenth century, indicating that the site and building has been a very busy one throughout history.

⑬ A149 road bridge over River Hun, Old Hunstanton

Ancient road bridge with foundations reputed to have been built by the Romans.

Few road travellers with an air of the mischievous about them will have resisted, at one point or another, quickening their speed before they drive over this venerable old bridge that takes cars, bicycles and careful pedestrians alike over the River Hun.

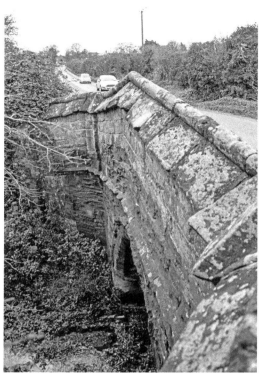

(Nigel Nudds)

It is, if anything, a landmark that isn't a landmark, a feature of the road that has been there, in one form or another, for hundreds of years – something that has long been taken for granted and crossed, more often than not, without thought or comment. Yet its walls are built upon stout foundations that are said to have their origins in Roman-occupied Britain, a fact that, if true, makes this humble little bridge one of the oldest known sites of a river crossing in the country.

The River Hun is a chalk stream approximately 3.7 miles long. Its source is in Hunstanton Park, from where it gently flows towards the North Sea and its mouth at nearby Holme-next-the-Sea. It is predominantly a seasonally running stream with its permanent flow only originating near to the bridge. Rivers that demonstrate this seasonal flow are known as winterbournes, which, as the name suggests, is a chalk or limestone stream that only flows in wet weather, predominantly the winter. Both the River Hun and the bridge that cross it are oft-ignored features of this stretch of the road and are certainly worthy of more recognition.

⑭ Holme-next-the-Sea

Small village close to the site of Seahenge, the famous Bronze Age timber circle.

Holme-next-the-Sea is a popular destination for bird watchers, with two adjoining nature reserves situated within the village boundaries, one owned by the Norfolk Wildlife Trust and the other

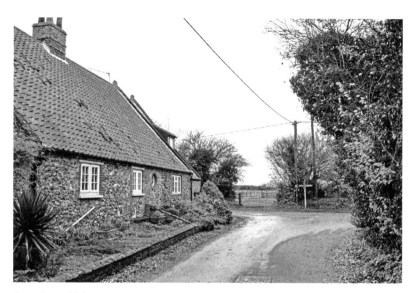

(Nigel Nudds)

by the Norfolk Ornithological Association. Its beach was the site of the prehistoric site of Seahenge, an extraordinarily well preserved timber circle with a large upturned tree root in its centre. Seahenge had been known and well regarded by locals for many decades and was largely left undisturbed, respected for its antiquity and perceived significance. All of those meant nothing to the team of archaeologists who eventually removed the entire structure, against the wishes of both the local people and Neopagans who believed, correctly in my opinion, that to move it from its original location was both a desecration and an insult to the religious beliefs of the original builders.

Seahenge may no longer be visible to the curious traveller but another religious site, the Parish Church of St Mary, most definitely is. Mentions of the site of the church date back to 1188, but the oldest part of the current structure is the tower, which dates from the fifteenth century. The main church building was demolished and rebuilt in 1888.

Holme-next-the-Sea's delights are not only favoured by 'twitchers' and plundering archaeologists; there is also plenty of appeal for the long-distance walker in and around the village as it's the site where the forty-six-mile-long Peddars Way meets with the Norfolk Coast Path, itself a sixty-three-mile walk that runs from Hunstanton in the west to Sea Palling in the east.

A coastal village with something for everyone. It's a pity that some of that had to be removed from where it belonged.

⑮ All Saints Church, Thornham
Church with Norman origins, whose construction was interrupted by the
Black Death.

The scourge of the Black Death in the fourteenth century reached its grisly peak in between the years of 1346 and 1353, with one of the numerous 'victims' of what was perhaps the worse pandemic in human history being the sudden cessation of work on Thornham Church in 1348.

The legacy of the darkest of times lurks in the village today with the 'spinney' (a small area of trees and bushes) in the village today known as the Plug Pits; however, this was once a burial ground for the village called the Plague Pits. Subtle.

There was further interruption into the building of All Saints (Norfolk's more modest version of Barcelona's world-famous Sagrada Familia, perhaps, on which construction has, 135 years later, yet to be completed) when the building of the church tower, commenced in 1633, was abandoned half-finished when the masons departed in some haste to help with the rebuilding of London after the Great Fire of 1666. Construction work on the tower was finally completed, in what is perhaps one of the great architectural examples of 'better late than never', in 1935, over 300 years after it had initially been abandoned to time and the elements.

No doubt King George V, with whose Silver Jubilee the completion of the tower was timed to coincide, would have been delighted to learn that his length of loyal service had, finally, been motivation enough to complete a project that had taken around 600 years from start to finish.

Among the church's striking features is its very wide nave, built with columns originally dating back to the thirteenth century, which

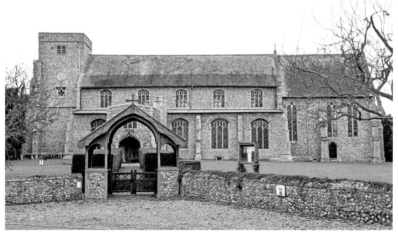

(Nigel Nudds)

were reset on higher bases during one of the church's many 'upgrades' in the fifteenth century. It is, as are many churches in Norfolk, a 'wool church', i.e., one whose construction was financed primarily by donations from rich merchants, particularly those who had benefited from the medieval wool trade. They enjoyed their prosperous lives while at the same time hoping that their generosity to the church would guarantee them a place in heaven after their death.

In a county renowned for its churches, All Saints Thornham is a delightful potpourri that is well worth a visit.

16 St Mary's Church, Titchwell
Exceedingly picturesque round towered church with small lead spirelet.

St Mary's Church in Titchwell is one of 124 of the round-tower type in Norfolk – more than in all the rest of England combined (in all, there are around 180).

Their proliferation in Norfolk is said to be due to the lack of local quarried stone that was suitable for the usual square foundation (whenever new stone was needed for work on the nearby church at Brancaster, locals did not hesitate to pillage the site of the village's ancient Roman Fort for any convenient pieces that might be useful!). Thus, these towers had to be built with rubble that was faced with the most commonly occurring rock – flint. The fact that there are very few quarries of note in the county add further credence to this theory.

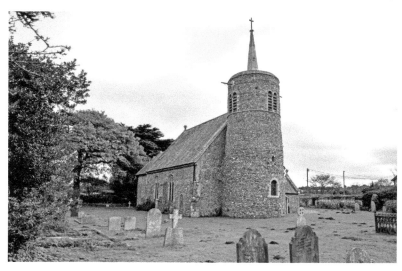

(Nigel Nudds)

St Mary's in Titchwell took on much of its current appearance during the fourteenth century, although substantial repairs to the building were carried out in 1844 and 1859, before a thorough restoration of the entire building was commissioned in 1890.

The lead spirelet on top of the tower is a special feature of this church and a striking form against the full, dark and heavy Norfolk skies on a winter's day. It was rather disparagingly described by noted twentieth-century historian Nikolaus Pevsner as 'a dear little lead spike'. It is topped by a large cross.

The interior of the church includes a Norman font. If its weathered appearance surprises the first-time visitor, then consider this: it previously spent several rather undistinguished centuries languishing in a nearby field being used as a drinking trough for horses! Among the glass in the church is a picture in the west window, which is of a sower setting out into the fields in spring time – a fitting tribute to the rich agricultural land that surrounds the church on three sides.

⑰ Village Cross, Titchwell
Medieval village cross.

Titchwell's medieval wayside or village cross is isolated but hauntingly prominent on the top of a small grassy mound at the eastern boundary of this small village, one that is much more famous for its internationally renowned RSPB bird reserve.

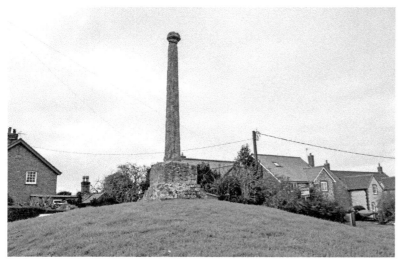

(Nigel Nudds)

It is a Grade II* listed building (defined as particularly important buildings of more than special interest) comprised of a socket stone, octagonal shaft and head. Very little else is known about the cross, which most likely served as a waymark and meeting point for walkers and possibly pilgrims. In regards to the latter, the shrine at Walsingham is approximately fourteen miles east of Titchwell, albeit on modern roads. Another purpose for these lonely sentinels was to warn unwary travellers that they were entering 'dangerous' territory, although Titchwell and its immediate environs seem unlikely to have ever fallen into this category.

In 1786, the Titchwell Enclosure Act was passed. This divided and enclosed the open fields, meadows, pastures and other common lands and grounds within the village.

As well as the well-known RSPB bird reserve, Titchwell boasts another, much less known, but utterly charming green space on Chalk Pit Lane. This is where the aforementioned pit, once used as a communal rubbish dump, was tidied up and reclaimed by the villagers as both a nature reserve and a place to visit if some quiet reflection and solitude are required.

18 Old Millstone, Brancaster
Hefty legacy of a village's milling past.

Two windmills once stood on the high and exposed ground to the south of Brancaster at the very top of the road that heads out of the village and inland towards Docking, which is, appropriately, known as Mill Hill.

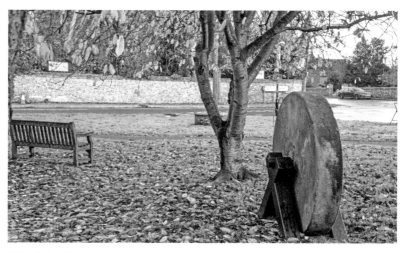

(Nigel Nudds)

The hill itself is a modest one that affords, nonetheless, a spectacular view of the village and coast as it crests at around 50 metres (approximately 160 feet) above sea level. So, while this is not quite the country of Noel Coward's Amanda ('Very flat, Norfolk'), it provided elevation enough for the village's two post mills (a windmill supported by a post, on which it pivots to catch the wind) to more than effectively do their jobs throughout the harvest season.

The two mills were known as the south and north post mills, with the former standing just a few yards south of its 'colleague' atop Mill Hill. It was an open trestle post mill with common sails and was turned into the wind via a tail pole. This mill ran one pair of 1.42 metre (4 feet 8 inches) stones and it is almost certainly one of these that is displayed on Brancaster's village green today.

Both mills were available for let or sale in 1829 along with, according to the *Norfolk Chronicle*, 'a good dwelling house with flour shop, bake house, granary, stables and a common right over the salt marshes at Brancaster.' It would seem that being Brancaster's version of 'Windy Miller' wasn't the most uncomfortable career option at the time.

For many years, the mill stone stood propped against the wall of the former JEC Powell Engineering works, often over time being lost inside the mass of tangled undergrowth that would regularly festoon it. With that old complex of buildings having recently being subjected to renovation or demolition, the mill stone was briefly hidden away from public view before being given its new home on the village green in 2016.

The large metal ring adjacent to the millstone would have been used by the village wheelwright for the making of wooden cartwheels.

⑲ Branodunum Roman Fort, Brancaster
Site of Roman Fort garrisoned by the Dalmatian Cavalry.

Although no visible physical evidence remains of this third-century fort, a visit to the site in 2012 from Channel 4's popular *Time Team* programme resulted in a great deal of new discoveries and insights into both the fort and its surrounding land, leading one member of the programme's team, Geophysics expert John Gator, to comment, 'I haven't enough superlatives to describe this site.'

You might wonder as to the reasoning behind that quote as you stand at the edge of the site, looking out over the gently rolling green meadow that was once the home to Rome's Dalmatian Cavalry or *Equites Dalmatae,* the Roman troops who would have been recruited from the part of Empire that now includes modern-day Croatia. Yet Branodunum

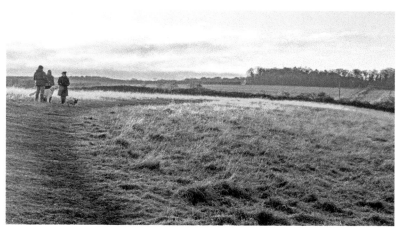

(Nigel Nudds)

was, as the programme discovered, an extremely significant site, one of nine such forts scattered along the English coast from Brancaster southwards towards Portus Adurni (Portchester Castle) in Hampshire.

It was most probably initially intended as a purpose-built garrison to guard the approach to the Wash, although it might also have taken on a central role in facilitating the movement of both people and resources throughout the entire region. With the fort's northern wall bordering the seashore, it would also have been an important focal point for travel and commerce, a fact illustrated by some of the rich discoveries (examples of which you would not normally expect to find at an assumed obscure outpost of empire) made by the *Time Team* archaeologists; items that included a rare pewter dish and ornamental belt decorations.

The site is managed by the National Trust and, while it is open to walkers, all forms of 'treasure hunting' on site, especially with metal detectors, is expressly forbidden. And rightly so.

㉒ AA Telephone Kiosk, Brancaster Staithe
One of only nineteen* remaining AA wooden sentry boxes left on or near public roads in Great Britain.

Of those nineteen AA call boxes left to endure the elements on Britain's highways, eight are Grade II listed, with the box situated on the layby to the west of Brancaster Staithe being one of those eight.

*According to www.theaa.com

29

That layby used to be part of the A149 road until the current stretch of road that is adjacent to it was constructed in the early 1970s. The straightening of that piece of road was done for reasons of safety, although, as one resident of the village pointed out to me, it has merely enabled motorists the opportunity to speed up as they leave the village on their way to Brancaster. Thus, that short stretch of tarmac upon which you might park and stroll upon to regard the box in question is, most undoubtedly, part of the original A149 as it once was, a gentle curve that eased the traveller in or out of the village of Brancaster Staithe.

The AA box was installed in the 1940s and is the oldest remaining of its type in Norfolk. It has the traditional upright 'sentry' box pattern with a gable above each face and a south-facing door. It still contains a (non-working) telephone and shelving.

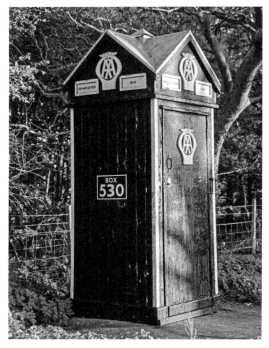

(Nigel Nudds)

㉑ Harbour, Brancaster Staithe
Popular harbour with the sailing fraternity that also sustains a local fishing industry.

This pretty harbour, with its juxtaposition of old and modern buildings dominating the skyline as you head towards the water's edge, is serene now – a popular destination and base for leisure users of all types, including sailors, canoeists, water skiers, naturalists and ramblers taking a break from their travels along the Norfolk Coastal Path.

Yet, Brancaster Staithe Harbour was once a very busy port that saw regular visits from sailing ships heavy with either coal or grain, with the latter intended for the village malt house that was, at its peak, said to be the biggest building of its type in England. Sadly, no traces of the malt house remain, while trade at the harbour started to decline towards the end of the nineteenth century. Happily, the harbour still supports a busy local fishing industry with mussels, cockles and crabs among the popular delicacies landed, sorted and sold here by the village fishermen.

The large central building that stands just behind the village green is Dial House. It was once a popular pub called the Victory, which

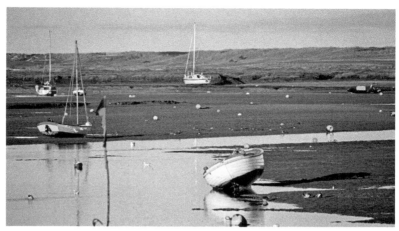

(Nigel Nudds)

closed in 1917. It subsequently became the home for many of the local wildlife wardens that have looked after the nearby wildlife reserve on Scolt Head Island. The building is now owned and managed by the National Trust, serving as the base of their popular Brancaster Activity Centre. After it was badly damaged by the floods that followed a tidal surge in 2013, the National Trust decided to completely refurbish the whole centre to incorporate both new flood mitigation technology as well as many new energy-efficient appliances. The refurbished building was officially declared open in July 2015.

There are few better places to go on a long summer's evening than Brancaster Staithe harbour when the tide is slowly ebbing up the creeks. It's an experience well worth taking a little time out for.

㉒ St Mary's Church, Burnham Deepdale
Serving the spiritual needs of the village since the eleventh century.

St Mary's at Burnham Deepdale, like its namesake at Titchwell a few miles along the road, is one of Norfolk's characteristic round tower churches, one that, in this case, dates to around the eleventh century. The tower, however, is believed to have been built at some point shortly after the Norman invasion in 1066, its construction showing those traditional Norman building techniques that involved using a mix of flint and rubble. This rather simple construction technique was initially thought to be preferred by local artisans as it rendered making the intricate corners needed in square towers unnecessary. It is now thought more likely that round tower churches were 'chosen'

31

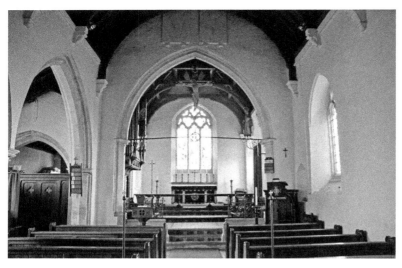

(Nigel Nudds)

for a cultural significance rather than that shape being easier to build if flint was the prime material used.

St Mary's is famous for its splendid Norman font, one that is square in shape, depicting in turn a figure that represents every month of the year. So, for example, someone is seen gathering in the harvest to represent September while the design that represents December shows a communal feasting scene to commemorate Christmas.

The bell in the tower that summons the local faithful to worship was made in the fourteenth century by local bell founder M. Derby of King's Lynn. The church also contains a lot of fine medieval glass, including the west window of the north aisle as well as the west window in the tower, which depicts an angel pulling on the chains of an incense burner.

The church has undergone three major restoration projects. The first, which was in 1797, saw the loss of the north aisle and south porch, with the latter being replaced during the third phase of restoration in 1898 along with the present north aisle. The church's origins, therefore, while definitely Saxon, have been subject to much alteration and change over the succeeding centuries, with the oldest parts of the building most likely being parts of the south and west walls.

Burnham Deepdale is one of the famous original 'seven Burnhams' of Norfolk, the others being Burnham Norton, Burnham Overy, Burnham Sutton, Burnham Thorpe (birthplace of Horatio Nelson), Burnham Ulph and Burnham Westgate. Modern-day Burnham Market, locally known as Chelsea-on-Sea, is the name given to the village that arose as a result of the 'merging' of Burnhams Sutton, Ulph and Westgate.

㉓ River Burn and Millpond, Burnham Overy Staithe
Old watermill, millpond and cottages.

As the A149 continues its easterly journey past the road that leads to Burnham Norton, it makes a gentle left-hand turn and heads temporarily northwards, passing as it does the old lower watermill on the right-hand side of the road, with the River Burn flowing under the road and into an exceedingly pretty millpond on the left.

The current watermill building was built in 1790 by Edmund Savory, a tenant on the Walpole estate at nearby Houghton Hall, where the great country house had recently been purpose built for the first British Prime Minister, Sir Robert Walpole. Savory had already been responsible for the mill bridge, used now as the crossing that takes the A149 over the River Burn.

Facing southwards onto the road and adjacent millpond is a row of terraced cottages that was once occupied by mill and farm workers. As time passed, the mill and adjoining buildings were developed to include eventually a malting (a building where cereal grain is converted into malt by soaking it in water), a granary, a blacksmith's, stables and cattle sheds with adjoining yards. The resident farmer was also responsible for around forty acres of prime agricultural land that surrounded the mill and outbuildings, making it a large and very prominent piece of real estate at that time, and certainly one of the more financially lucrative in the area.

By the end of the nineteenth century, the waterwheel drove three pairs of milling stones with its river-sourced power now supplemented by a 16-hp condensing steam beam engine that drove a further four sets of stones. The waterwheel used at that time was still intact into the mid-1960s.

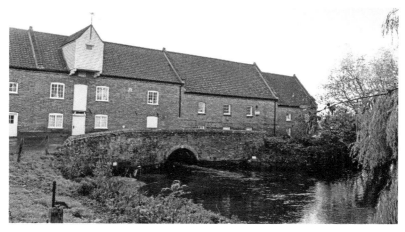

(Nigel Nudds)

The River Burn, which is also known locally as 'Nelson's River', is 9.8 miles long with its source in the village of South Creake. Burnham Overy lower mill is the second of two watermills on its route, the first being Union mill a short way upriver. It was one of only three water mills in Norfolk to use both water and wind power on the same site, the other two being at Little Cressingham and Thurning.

24 Windmill, Burnham Overy Staithe
Grade II* listed building tower mill.

Burnham Overy Staithe windmill was built in 1816 for Edmund Savory, the miller who was already managing the nearby watermill, which was known as the Lower Mill, over the River Burn.

This impressive windmill, one that first captured my imagination when I was a small boy and has enchanted me ever since, is a six-storey tower mill, i.e. a windmill built of either stone or brick. A wooden roof or cap is then sited on top of the tower, which can be rotated to bring the sails into the wind. Burnham Overy Staithe windmill has a fantail plus four double patent (a combination of spring-adjusted shutters on the sail with strong canvas strips) sails attached to a cast-iron wind shaft that powered three pairs of millstones.

Edmund Savory worked both mills until his death in 1863. The mills then passed onto his son John, who was, it seems, not quite so respectable

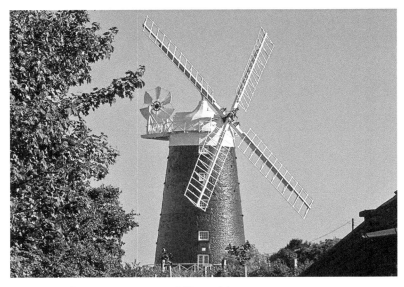

(Mark Oakden at www.tournorfolk.co.uk)

or even as hard working as his father. In 1870 and 1873, John Savory junior was fined £5 (equivalent to around £700 today) for owning an unlicensed horse (a law had been passed in 1869 declaring that all working horses had to have a licence). Savory, out of pocket, carried on working at the site until 1888 when, doubtless keen for a less arduous (and potentially expensive) means of making a living, he put the whole site up for sale by auction. Much to, I suspect, his intense disappointment, it remained unsold and he continued to work there until 1900.

In 1973 the windmill briefly recreated the scene that would have been familiar to many during its working life when, during a strong gale, its skeleton (frame only) sails were said to have briefly rotated in the wind.

In 2006 the windmill commenced a new part of its life as a residential holiday let.

25 Harbour, Burnham Overy Staithe
Popular harbour for leisure users, which once served as an important port.

A hive of colourful activity will normally greet any casual visitor taking a stroll by the water's edge as high tide approaches at this pretty harbour, a happy haven whether you are a keen sailor, walker, bird watcher or, quite simply, someone who, like me, just likes to sit and watch the world float by.

Burnham Overy Staithe harbour may be tranquil now but that was not always the case. It used to be an important port for the surrounding villages along this stretch of coast and was, with Brancaster Staithe a few miles along the road to the west, one that saw a lot of commercial shipping either land its wares here or make its way up the River Burn to Burnham Overy Town. The commercial importance of the

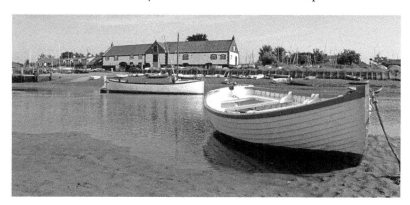

(Mike Carroll)

harbour began to ebb with the arrival of the railways at the Burnhams (Burnham station was opened in 1866, renamed Burnham Market in 1883 and closed in 1952) with the last cargo believed to have been delivered and unloaded at the Staithe here just before 1920.

To the east of the main tidal creek at Burnham Overy, the former salt marshes that lay between the beach and agricultural land have been reclaimed to form fresh water meadows that now make up part of the Holkham estate. There is a footpath just under two miles in length that links Burnham Overy Staithe to the beach, which runs along the top of the embankment (bank), serving to protect the current fresh water marsh from the tidal creek.

26 Beach, Holkham
Beach renowned for its big skies and movie connections.

If anyone has ever told you that Norfolk is famous for its 'big skies' and you want to get a clearer idea of what they are talking about, then take a trip down to Holkham beach via the narrow roadway that will take you there from the A149 on the western outskirts of the village.

Once you are on the beach and standing on the wide-open expanses of sand that wait for you there, you will know what the term 'a Norfolk sky' is all about. A vast canvas awaits that seems to go on forever, with you, a small and insignificant dot, stood underneath with nothing better to do than admire its magnificence and wonder why the equally magnificent beach that you are stood upon isn't as well-known as some of the more famous beaches in the world such as Rio, Bondi, Source d'Argent …

Holkham? Well, why not?

In truth, Holkham is the superstar beach that doesn't need the headlines or glamour that accompany such locations. And it certainly doesn't, and will never need, the associated and occasionally crass commercialisation that goes with such things. Holkham's emptiness and loneliness are what gives it the wild beauty that has been noted by the film industry, with perhaps its best-known starring role being alongside Gwyneth Paltrow, who strode along it in the final scenes of 1998's *Shakespeare In Love*.

Holkham is also a haven for wildlife, with the beach backing onto an area of sand dunes, salt marsh and a large belt of pine

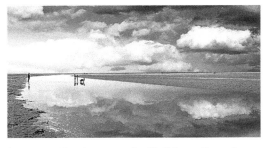

(Laurane Herrieven at the Holkham Estate)

trees known as the Holkham Pines. These all seamlessly unite to form a number of habitats for rare animals and plants, with the whole area part of a nature reserve that covers eleven miles of this part of the Norfolk coast.

In 2015, over 100 travel writers and editors were named to ask their favourite UK beach by a well-known travel website. The beach that came out on top, and convincingly, was Holkham.

27 Wall, Holkham Hall
Stern red-faced guardian of Holkham Hall Park.

My late father would never tire of telling me, whenever we drove past the outskirts of the Holkham Hall Park, that its all encircling wall, 'could clearly be seen from space'. In truth, of course, the park's boundary wall cannot be seen from space; indeed it all but disappears at a height of just a few hundred feet up, but it is, nevertheless, an impressive sight and a tribute to those Victorian craftsmen who designed and built it between the years of 1833 and 1839.

Even allowing for natural breaks for entrances, gateways and other necessary gaps, the wall is still nine miles long, enclosing a park that has many points of interest for the curious explorer to investigate, not least the famous Monument. This is a 120-foot-high Corinthian column that was erected in memory to Coke of Norfolk, the 1st Earl of Leicester, and completed in 1850. The column is topped with a wheat sheaf while the square plinth at its base is decorated with panels and carvings that depict various aspects of Coke's contribution to agriculture. It was designed by

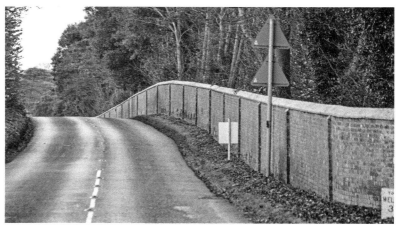

(Nigel Nudds)

the early nineteenth-century English architect William Donthorne and paid for by public subscription. Donthorne was also responsible for the design of three great Norfolk houses, namely Elmham, Hillington and Watlington Halls. But don't think about going to visit them as they have all long since been demolished! You can, however, visit another of his great works, namely Highcliffe Castle in Dorset.

Another interesting point of reference within the park is the ice house, which is believed to have been built in the middle of the eighteenth century. It would have been used to store ice throughout the year, the natural ice and snow of winter being taken into the ice house and packed tightly with insulation materials such as straw or sawdust, ensuring that it remained frozen for many months.

The park has designated walks for visitors that allows them to explore the 3,000-acre grounds at their leisure. There is also, for the more athletically minded, a ten-mile cycle route that passes through the park grounds. It is also the home to a large herd of fallow deer, as well as a smaller one of red deer. They are fairly oblivious to visitors and you will be able to walk past and observe them quite effortlessly.

28 Holkham Hall
Eighteenth-century country house constructed in the Palladian style for Thomas Coke, 1st Earl of Leicester.

Holkham Hall is frequently described as one of England's finest examples of the 'Palladian revival style of architecture' – the said style being influenced by the ornate European style of great building that was favoured and inspired by some of the designs of Andrea Palladio, the sixteenth-century architect from Venice.

The modern-day Holkham Hall is the privately owned and lived-in family home to the 8th Earl of Leicester. Parts of it are open to the

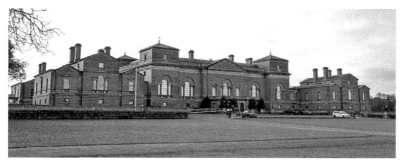

(Nigel Nudds)

public, including the Statue Gallery, designed with the intention of displaying classical statuary in niches within the room that were specifically built to the dimensions of the statues that would be placed in them. One amusing story relating to this part of the house tells of the delicate lady visitor who, in 1772, admitted her admiration for the statues, but added that it was a little impractical to admire the art too closely as the statues were rather scantily clad. The Victorian custodians of the collection took note and commenced the use of strategically placed shrouds in order to preserve the sensitivities of their visitors. Luckily for us now, no such restrictions apply and the collection, reckoned to be the most complete of any private house in England, can be seen as nature or, rather, the sculptor intended.

Thomas Coke, the 1st Earl of Leicester for whom Holkham Hall was built, has been widely attributed as the pioneer behind four-year crop rotation, although the concept and introduction of same is now credited to a group of Norfolk farmers including the famous Turnip Townsend from nearby Raynham. Coke is now more regarded as the man who promoted and encouraged the widespread adoption of crop rotation rather than being its inventor himself.

Admire the house and park (the designers of which included Capability Brown) by all means but, as you enjoy your visit, spare a thought for the men of Holkham and the surrounding farms who developed and introduced a farming practice that was adopted all over the world – a triumph of ingenuity that is as long lasting and impressive as the great house that forms the estate's centrepiece.

㉙ Site of Railway Bridge, Wells-next-the-Sea
Railway bridge remains at turnoff to Wells-next-the-Sea.

As the A149 rolls down to the popular coastal town of Wells-next-the-Sea at the junction with the B1105 (which wends its way south to Fakenham), one cannot help but notice the large stone edifices that flank either side of that road – the remains of the railway bridge over which passengers would travel while en route from Holkham to Wells stations.

It must have been quite a sight in its time, that of a train labouring over the bridge above the road to Wells, creating a potent mix on some days, no doubt, of steam and coastal mist that might have made as good a subject for a Turner painting as his railway masterpiece from 1844, *Rain, Steam and Speed*. Sadly, few images of the bridge in its prime exist, as much a memory now as the stretch of railway that once served this part of the coast.

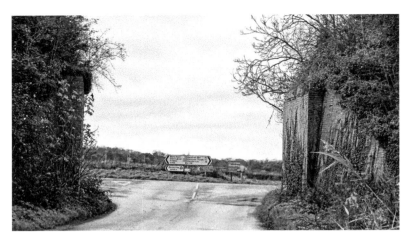

(Nigel Nudds)

The old railway station at Wells-next-the-Sea is now the Old Station Pottery and Bookshop, which is on the corner of Maryland and Polka Road in the town. It was first opened in 1857 by the Wells & Fakenham Railway (running from Wells up to Fakenham), later becoming part of the Great Eastern Railway's Wymondham–Wells branch. It then became a junction in 1866 with the arrival of the West Norfolk Junction Railway, which ran from Wells to Heacham.

It's hard to believe that modern-day Norfolk, which now contains just thirty-two working railway stations in total, used to boast an additional 139, many of which fell under the merciless consequences of the Beeching cuts. Passenger services between Wells and Heacham were closed from June 1952, although the line initially remained open for freight traffic. However, just one year later, the damage caused to the track between Wells and Holkham was so considerable that British Railways decided there and then to permanently close that stretch of line. Wells station continued operating a service to and from Dereham until the station at Wells was finally closed to passenger traffic in October 1954.

30 Quay, Wells-next-the-Sea
'All I ask is a tall ship and a star to steer her by'.

For me, the quayside at Wells-next-the-Sea is exactly what one should look like.

It's near perfect.

Ships of all shapes and sizes can sail right up into the town centre and berth a short stroll away from all that a weary, hungry or thirsty sailor would ever need. The romance is right there and the excitement

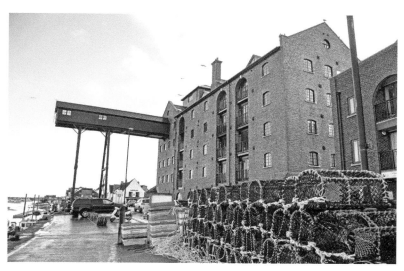

(Nigel Nudds)

of seeing what ships may be berthed in the town is one that remains with me to this day.

During the Great Flood of 1953, a large ship didn't just find its berth at the quayside; it ended up on the quay itself with the sheer force of the tidal surge floating the 160-ton Sea Scouts training vessel MTB (Motor Torpedo Boat – clearly these were Sea Scouts who really were prepared for anything!) *Terra Nova* up and onto the dockside where she remained, stranded, after the flood waters had ebbed away. A popular pastime all year around in Wells-next-the-Sea, and one you may feel inclined to do yourself (and it is thoroughly enjoyable on a sunny day) during your visit, is to sit and eat your fish and chips on the low wall that separates the quayside and its vehicles from the coast-hugging road as it threads through the town.

Take time, then, to think of the chaos, the noise and the fear that would have been in people's minds that night as the flood waters surged over the top of the quayside. Wells-next-the-Sea is an exceedingly pleasant little town now, a small but hard-working port that has had, in recent years, to increasingly re-invent itself as a tourist destination, something that it has done without losing its raw and suitably salty character.

One very familiar landmark on the seafront is the granary with its distinctive overhanging gantry that stretches across the road towards the quay. Construction on this working part of the town started in 1902 and took two years to complete, with the granary finally ceasing its legitimate 'duties' in 1990. It has since been converted into holiday apartments with the very top of the granary now amusingly marketed as 'The Granary Penthouse'.

31 Stiffkey Fen
Nature reserve with reed bed and fresh water lagoon.

Stiffkey Fen nature reserve covers approximately thirty-five acres of reed beds and fresh water lagoon, created from farmland by Lord Buxton, a noted naturalist and philanthropist who, as well as being a director of Anglia Television, was also responsible for the creation of that channel's famous *Survival* series.

With help and support from the Environment Agency, the wetland was improved by slowing the flow of water through the fen. This meant that the ecosystem developing there was able to settle down and mature without the regular disturbance and erosive effects of fast-flowing water.

The fen is hugely popular with both birds and birdwatchers. Among the feathered winter visitors is a roost of up to 4,000 northern lapwings (or Peewit) as well as thriving colonies of water rails and bearded tits.

Parking for Stiffkey Fen is available just up the hill to the west of the fen. The careful walker then crosses to the inland side of the road where a hedge-lined path has been provided.

The village of Stiffkey (pronounced locally as 'Stewkey' – Norfolk enjoys place names which are pronounced entirely differently to how an outsider would expect. Other examples include 'Hunston' for Hunstanton, 'Windham' for Wymondham and 'Haysbro' for Happisburgh) was the parish whose rector, one Harold Davidson, found himself facing charges of immorality in 1932 that led to him being defrocked – a wonderful but ultimately tragic tale to which Nick Conrad refers in his foreword. Despite Davidson's transgression, he remained a popular figure in the area and, upon his death (he was

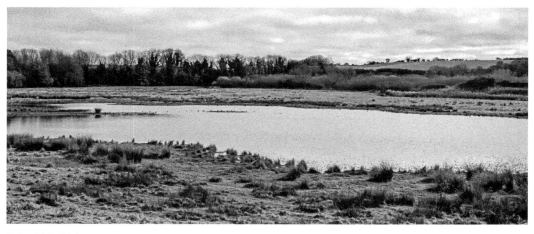

(Nigel Nudds)

killed in Skegness by a lion, and if that doesn't make you want to find out more about him, then nothing will), the villagers of Stiffkey asked his family if he could 'come home' and be buried in Stiffkey rather than being interned in the family crypt in Hampshire.

His grave, in the churchyard of St John & St Mary in the village, is tended by the residents of Stiffkey to this day.

32 Morston Quay, Morston
Part of the Blakeney Nature Reserve.

Situated within the protective influence of the famous reserve at Blakeney Point, Morston Quay forms a wonderfully idyllic canvas on any day of the year with its rich mix of those characteristic big Norfolk skies and ever changing salt marshes. Happily, the quay has not totally surrendered itself to the tourist trade and still supports a modest fishing industry as well as providing a regular supply of shellfish to local restaurants and pubs, with the crabs and lobsters being particularly highly valued catches in this part of the world.

Morston Quay is also the base for the popular daily boat trips out to see the seals and admire the views in and around Blakeney Point itself. This service was started up in the mid-1930s by W. J. Bean and his brother E. A. K. Bean. After a brief hiatus during the Second World War, the service was resumed in 1946 with W. J. Bean's two sons joining the business as soon as they left school in the late 1950s. The Bean family continue to run trips to see the seals to this day.

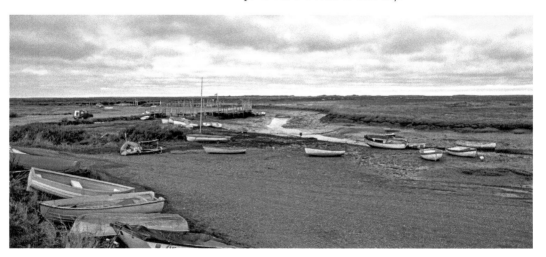

(Nigel Nudds)

The gentle pace of life and work around Morston now belies the fact that the village would have been a major port right up until the seventeenth century. Like so many along this part of the coast, it saw usage decline due to a combination of factors that included the rise of the railways and the silting up of many of the shallow channels and creeks along this part of the coast, which made navigation virtually impossible to all but the smallest of craft.

Morston Hall, a nearby country house built in the seventeenth century, is now a four-star hotel and restaurant.

㉝ Mariners Hill, Blakeney
Artificial lookout point for harbour.

Little is known about Mariners Hill, the large grass-covered mound that gently rises to a height of around 25 feet from its inland site close to the village harbour. It was once thought to possibly be a motte (a flat-topped mound of earth, often made artificially in the Middle Ages, onto which was built a wooden or stone defensive structure). However, there are no available historical records to substantiate this and neither is there any physical evidence to suggest that a defensive wall (bailey) was ever built on the site.

Thus, however romantic the notion might be that Mariners Hill is the site of a modest castle that acted as defensive hub for Blakeney in the Middle Ages, it is more likely to have been a raised point created

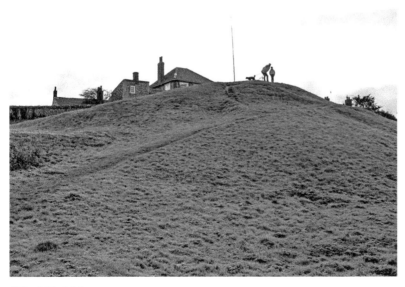

(Nigel Nudds)

44

much later to act as a lookout over the harbour and surrounding area, as well as to provide a site for a beacon.

Blakeney itself, like so many other settlements on this part of the coast, would have grown up and prospered as a busy medieval port that would, like the harbour at nearby Morston, have been used less and less as the creeks and channels began to silt up. The village itself is exceedingly pleasant, consisting of many attractive buildings as well as a guildhall that dates from the fourteenth century and contains an early example of a brick-built vaulted ceiling.

A very agreeable village to while away an hour or three as you go about your travels.

34 Church of St Nicholas, Blakeney
Mighty Gothic church that reflects village's historical importance.

The impressive church of St Nicholas in Blakeney was founded in the thirteenth century. However, most of the building that is visible today dates from the fifteenth century, with its scale and grandeur reflecting the importance of Blakeney as a seaport at that time.

A standout feature of this church is its second tower at its eastern end, the top of which was used as a site for a beacon. This raises the interesting possibility that Blakeney either had two sites used for beacons or, alternatively, that the nearby Mariners Hill had been proposed and constructed, but was then never used as an alternative site.

The church, which has a Grade I listing due to its exceptional architectural interest, contains several significant features within its

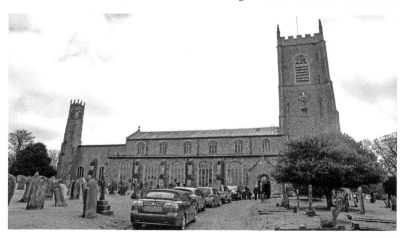

(Nigel Nudds)

interior, including the vaulted chancel with its lancet (slender and pointed arched) window and the hammer beam roof in the nave. With the church being as prominent as it is (and was), it was unsurprisingly an easy target for Henry VIII's 'reforms' in the sixteenth century. Much of what was lost to the architectural vandalism of that time was restored in the eighteenth century by the Victorians, who perceived that falling attendances at church would be reversed if the populace had grander and more ornate buildings in which they could worship. This approach was a fairly widespread one at the time, borne out by estimated figures provided by the Church of England that claim around 80 per cent of their ancient buildings were affected by this misguided 'enthusiasm' in one way or the other, with their collective efforts ranging from minor internal changes to complete demolition and rebuilding. (However, the Victorians did not always appreciate the concept of conservation and appreciation when it came to antiquities. At Stonehenge, visitors were given hammers and chisels and invited to hack pieces off the famous stones that they could then take home with them!)

Luckily, Blakeney did not suffer too much at the hands of the Victorian enthusiasts. The building was resurfaced and repaired to a fairly decent standard throughout, while its new woodwork was designed to match the surviving older pieces as closely as possible. Thankfully, their efforts did not extend to erasing any of the pre-Reformation graffiti that the church contains today, including at least thirty different depictions of ships – testament again to the village's importance as a medieval shipping hub.

35 Marshes, Cley-next-the-Sea
400 acres of unspoilt marsh managed by the Norfolk Wildlife Trust.

The Norfolk Wildlife Trust (NWT), which purchased the 400 acres of marsh they now manage at Cley in 1926, did so with the express objective that the land would be held 'in perpetuity' as a bird breeding sanctuary. With the site now less than a decade away from its centenary, the work they have done, and the progress that has been made in developing this site as one of the most best-known nature reserves in the region, should be considered as nothing less than an absolute triumph.

One of the reasons for this celebration is that the methodology that the NWT have adapted and developed in that time went on to provide a seminal blueprint for nature conservation that has since been replicated across the UK. Thus, while this wonderful site may not be as well-known as those at the UK's 'showcase' locations,

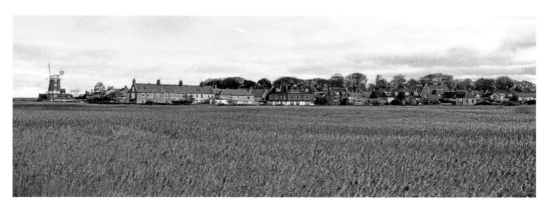

(Nigel Nudds)

e.g. the WWT site at Slimbridge or the RSPB one at Minsmere, both of those will almost certainly have been influenced, one way or another, by what was first done and achieved at Cley.

The site includes a shingle beach, saline (salty) lagoons along with marsh (also used for grazing) and extensive areas of reed bed. Wintering and migrating wildfowl that can be seen at Cley include the marsh harrier and bearded tit, as well as the elusive bittern whose booming cry is one of the great iconic sounds of Norfolk's marshes. A new visitor centre was opened at the site in 2007, one that was designed and built to be as eco-friendly as possible. It's a very unobtrusive building that 'hugs' the ground in order to minimise its visual impact on the site's surroundings – its roof being covered in sedum-moss (which thrives in the coastal air) in order to give the building a green 'footprint', one that not only attracts butterflies and other insects, but also helps with the absorption and discharge of rainwater into the site's drainage system.

Even if you do not have time to visit the marsh, do take a moment to admire the view from the visitor centre across to the marshes and sea. It is described on the reserve's website as 'breathtaking', which is a more than fair description for a view as good as any you will find along the A149.

36 Windmill, Cley-next-the-Sea
One of the iconic sights of this stretch of coast.

The sight of Cley windmill standing proudly against a (usually!) deep-blue Norfolk sky remains one of my favourite images in the entire county. I rather think it must also be one of the most painted and photographed.

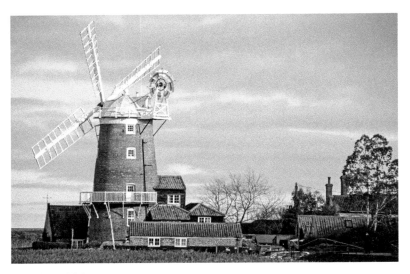

(Nigel Nudds)

It was built in the early part of the nineteenth century, duly being brought to the attention of the world at large when, in 1819, an advertisement in the *Norfolk Chronicle* outlined that the mill, 'newly erected' and owned by a family of the name of Farthing, was for sale. Unfortunately for the Farthings, their attempt to sell this lovely windmill garnered the same response as that met by a certain John Savory Jnr (see 24) and failed to sell, remaining in the family until the matriarch of the family, Dorothy Farthing, died in 1875.

Its time as a working mill was, by then, already running short and, after being worked by its new owners until 1912, it was eventually bought by a Mrs Sarah Wilson who, in an act of great foresight (and for just £350 – in 1996, the windmill was put up for sale with an asking price of £1,500,000), converted the mill to a holiday home. This meant that all of the working machinery had to be removed, with the gear wheels being cut in half and used as decorations within the newly refurbished mill.

Being, as you will see, in a very exposed location, the mill has been subject to quite a battering from the elements over the years. It was flooded to a depth of around eight feet during the Great Flood of 1953 and, seven years later, it was determined that the sails needed to be replaced – this was done at a cost of £1,500.

For those with technical inclinations, Cley is a five-storey tower mill with a dome-shaped cap and gallery. The cap is now fixed and immovable. The four sails have a span of seventy feet and are carried on stocks fifty-six feet long. They were, in the nineteenth century, able to provide enough power for three pairs of millstones.

48

The mill was listed for sale again in 1982 but, again, failed to attract a suitable bid and, a year later, planning permission was sought and secured to convert the mill and its adjoining buildings into a guest house with separate self-catering units. It remains a high class and very popular guesthouse to this day, one full of character, and available, if you wish, for both private house parties and weddings.

37 St Margaret's Church, Cley-next-the-Sea

Impressive church in the perpendicular style (an early English architectural style that favoured, among other features, pointed arches or lancets. You will now be able to impress friends and family alike by pointing these features out and referring to them as being characteristically perpendicular).

St Margaret's in Cley always strikes me, like so many Norfolk churches, as one of those that, were it, for example, in Paris, Rome or Florence, would be venerated all over the world and be one of the great tourist attractions of those cities. But, no, it is content to serve as the Anglican parish church for the village and has been since the early part of the fourteenth century, with its first major piece of renovation coming towards the end of that decade when a large south porch was added to the existing building.

Inside and out, this is a spectacular church, with an air of ancient magnificence intensified by the derelict north and south transepts and a spacious, echoing nave. It sits, not unlike Ely Cathedral some

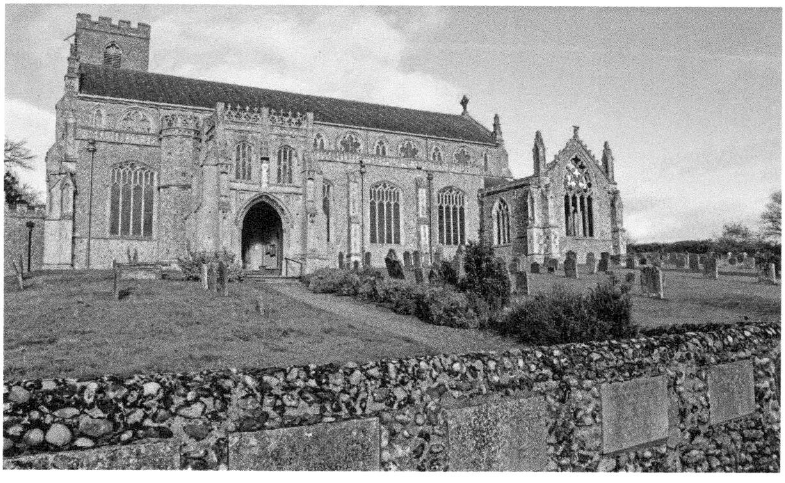

(Nigel Nudds)

sixty miles to the south, among the flatness of fields and marsh like a mighty ocean liner – a building that demands just a little awe and respect. It certainly has mine. This was an objective, you suspect, that was in the mind of its architect back in the early 1400s when it was conceived and began its life. A building to impress its congregation.

The church became a popular gathering point in 2008 when a rare white crowned sparrow (a bird native to the United States and seen very infrequently in the UK) was spotted in Cley. The flocks of excited twitchers that made their way to Cley donated over £3,000 to a collection for the church's restoration during their brief stopovers in the village, with the bird being thanked by way of having its image included on a window in the church.

38 Heath, Salthouse
Popular destination for geology students!

I've long heard (and tired of, quite frankly) the argument that Norfolk 'doesn't have any geology'.

What nonsense.

If of course, that means the sort of geology that consists of spectacular mountain ranges, outcrops of bare, sheer rock, sculpted limestone pavements, networks of caves or remnants of a violent volcanic past, then yes, the region is not exactly a hub for all things igneous and unyielding. But to say there is 'no geology'?

Salthouse Heath forms part of the Cromer Ridge of old glacial moraines (an accumulation of glacial debris, primarily soil and rock that has gathered at the edge of a glacier) that stand along the part of Norfolk coast that includes Salthouse. The area thus has the 'honour' of

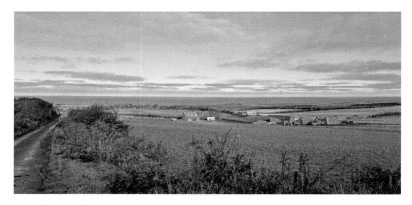

(Nigel Nudds)

being the front line of the last great ice sheet to have covered the UK around 10,000 years ago – with these gentle ridges being the end result.

So much for 'doesn't have any geology'. Jurassic Park it is not, but if a visit to Quaternary Park is what you are looking for, then this part of Norfolk, the Salthouse Heath included, is where you need to come. It is a hugely significant and important landscape feature – one that has been recognised as such throughout history, as it also contains, for example, a large cluster of Bronze Age mounds.

The ridge, of which the heath is part, contains one of the highest points in East Anglia, reaching a maximum height of 335 feet (around 1.15 per cent of the height of Mount Everest!) along its near nine-mile length. The predominant vegetation on the ridge is gorse and heather, a natural feature of such soils and landscapes that has, in recent years, been encroached upon by invasive birch and oak woodland. In early summer, it is, so I am advised, a good place to hear nightingales and nightjars – an aural treat if ever there was one.

You could be standing at the very edge of the last great glacier's advance northwards. Think about it. Doesn't that give you goose bumps?

39 St Nicholas Church, Salthouse
The church with the smiling lions.

St Nicholas Church in Salthouse can trace its origins, like so many Norfolk churches, back to the thirteenth century, although the handsome church that stands, sentinel like, today is a 'new build' – at least, if you consider that the work undertaken to rebuild it by a Sir Henry Heydon in

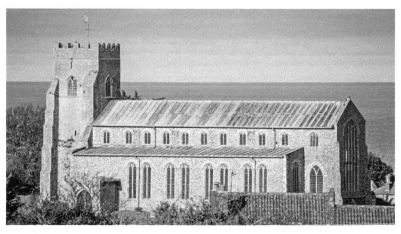

(Nigel Nudds)

the fifteenth century counts as such. It is a handsome church. It doesn't have the same ability to make you stop in your tracks as its mighty namesake in Blakeney might but, then again, this church does have much more of a 'come on in and have a chat to God' quality about it.

The church is another example of the late perpendicular style (remember the pointed arches? Plenty of them to see here) with an interior that has been designed to let in as much natural light as possible; the great east window is a particularly striking feature. Much of the tiled floor is original, while the octagonal font, carved at the time the church was rebuilt, contains some exceedingly friendly lions on four of its edges – fitting really for a church that is clearly used and loved, and that exudes approachability.

40 Muckleburgh Collection, Muckleburgh
Site of largest privately owned military museum in the UK.

Fittingly, the Muckleburgh Collection, a military museum that was opened to the public in 1988, is sited on a former military base called the Weybourne Anti-Aircraft Training Camp. Originally known as Carvel Farm, the site was first used in 1935 by the Territorial Army as a temporary hub for its summertime operations before, two years later, and with the threat of war now a very considerable one, the decision was made to make the camp a permanent one. This led to many permanent structures and defences being built, much, you suspect, to the concern of the local populace whose fears of another global conflict would have been considerably heightened by the sudden frenzy of military activity at the 'old farm'.

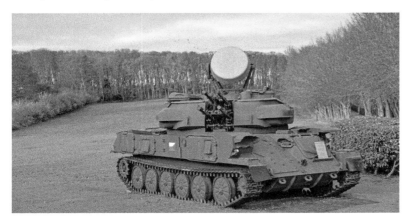

(Nigel Nudds)

The camp was certainly regarded as having enough strategic importance during the Second World War to have been surrounded by an anti-tank ditch as well as defences that included barbed wire and numerous gun emplacements, with the lonely clifftop that overlooked the North Sea dotted with anti-aircraft guns, batteries, trenches and pillboxes. It was even deemed important enough for Winston Churchill to visit the site in 1941 to witness the demonstration of a new anti-aircraft weapon.

As the war progressed, the infrastructure within the camp became ever more complex (and secret!) with, ultimately, a grass airstrip being laid out that can still be used to this day. This is sobering evidence of just how seriously the threat of an invasion by Nazi forces was taken along this whole stretch of coast, which is still littered with a vast pillbox legacy.

The museum that occupies the site today takes up around 300 acres and is a popular set location for television films and documentaries.

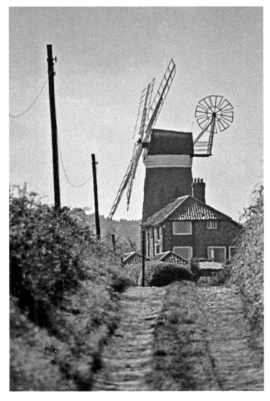

(Nigel Nudds)

41 Windmill-Weybourne
Five-storey red-brick tower mill.

One of the endearing qualities about Weybourne's striking windmill is how it appears exactly like a child's drawing of a windmill come to life. It has an almost fairy-tale quality and would not look out of place in an illustration for one of Hans Christian Anderson's stories of magic folk.

Four sails, four windows in the tower and a cottage attached. Isn't that what all windmills should look like?

It was, of course, a working mill, one that was built in 1850, which when it was in operation powered three pairs of millstones that were located on the windmill's second floor. It was worked until 1916 when, in accordance with both local and national priorities overseas, it fell into a state of disrepair and was neglected for nearly a decade until some restorative work was attempted. Unfortunately, this meant the mill losing most of its working parts aside from the wind shaft (the shaft driven by a windmill's sails). By 1929, the windmill had passed into the hands of a new owner and was now bereft of all its sails, the fantail and gallery – a sad sight indeed.

The windmill was purchased privately in 1967, after which some serious and committed restoration work followed. This continued into the next decade with the assistance of a grant from the Norfolk Windmills Trust, which was used to restore the windmill's paintwork. A second grant was awarded in 1973 for further restoration and maintenance work before, in 1982, the 'all new' windmill was again put up for sale and promptly sold for £125,000 (equivalent to a little under £450,000 today) to a private buyer.

Weybourne windmill is not open to the public.

42 Park, Sheringham
Landscape park and gardens.

You may not have heard of Humphry Repton. He is regarded as the last of the great English landscape designers, a successor, almost, to the internationally renowned Capability Brown – he who had some input at the park in which the opulent Holkham Hall stands (see No. 28).

Repton, the designer behind Sheringham Park (it is worth the short detour off the A149 to pay a visit) was astute enough to spot a gap in the market upon Brown's death in 1783, noting that, with his demise, there was no obvious successor in a trade that had not only made Brown comparatively wealthy and famous, but, more to the point, had led to him becoming a bit of an eighteenth-century celebrity – the Alan Titchmarsh of his day. Unlike Brown, however,

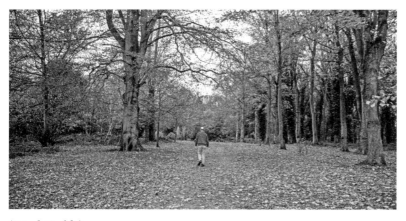

(Nigel Nudds)

Repton had not dedicated his working life to the landscape but had worked in the textile industry as well as tinkering as a journalist, dramatist, artist, political agent and even as a secretary-cum-companion to William Windham, the Whig statesman who resided at nearby Felbrigg Hall.

Not too unreasonably, having shown a little flair for artistic pursuits, Repton reckoned he could turn that to landscaping and the creation of the large and spectacular gardens that had been the preserve of Brown. He duly took it upon himself to market his 'new' skills by networking with as many of his wealthy and influential friends as he could, hoping that one of them would award him a commission, his applications being backed by a collection of what he called his 'red books'. These were lavishly bound and illustrated volumes of his works and proposals, an ostentatious illustration of his ego as well as his plans for contemporary landscape design.

Repton's mix of ambition, charm and chutzpah was enough to get him the commission to design Sheringham Park, winning over its owners, Abbot and Charlotte Upcher, in 1812 with his ideas for their estate.

The park today boasts a rich variety of trees and shrubs, in particular a number of different rhododendrons as well as magnolia, maples and the exquisite snowdrop tree, all within a landscape designed by a rather remarkable man.

43 Sheringham
Mare Ditat Pinusque Decorat – *'The sea enriches and the pine adorns'.*

A fishing town.

Sheringham, unlike Heacham with its village-based brick industry (see No. 9), prospered with the coming of the railways, which made it all the easier for the locally caught fish to be transported to local markets and beyond. Throughout the 1900s, the town's 'core' catch of crabs, lobsters and whelks were in demand as far afield as London. You only need to sample any one of them to discover for yourself why they were, and remain, so popular.

The town's great period of growth commenced when the Midland & Great Northern Railway arrived in the town (see No. 44), this coinciding with the construction of many of its fine and, occasionally, stand-out buildings, with flint being the major material used. This meant that new builds in the town during the late nineteenth and early twentieth centuries had to be designed and built with the limitations of flint in mind, a building style that is known as the vernacular. The town's

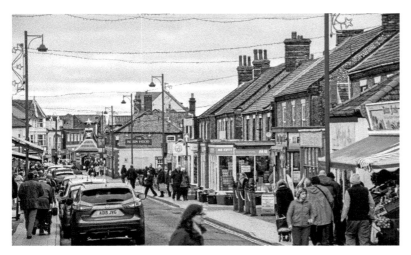

(Nigel Nudds)

cinema, constructed in 1914, is a fine example of this technique; however, even considering those limitations of the locally available materials, you cannot, upon seeing the building today, mistake its original purpose as being anything but a cinema – something that lends great credit to both architects and builders alike.

Sheringham is, along with its sister town Cromer, famous for its lifeboats and the legacy the brave men and women who served on them have given the town. Four of Sheringham's original lifeboats still exist, with three of them owned by the Sheringham Museum Trust. Launching any one of the lifeboats is no easy task for, with the town lacking a harbour to call its own, they must be launched from the beach and towed into place by a tractor.

Norfolk's famous 'Singing Postman', aka Allan Smethurst, although born in Lancashire, was brought up in the town – as good a place as any to grow up in and learn all about Bishy-barney-bees and Dickeys ...

If Norwich is a 'fine city', then Sheringham is a 'gret ole' town ...

44 The Poppy Line, Sheringham Station
All aboard ...

When the Midland & Great Northern Railway opened Sheringham station in 1887, they must have had a fair old inclination of what lay ahead for the town, for at that time the station stood in the open countryside with Sheringham itself no more than a small fishing village.

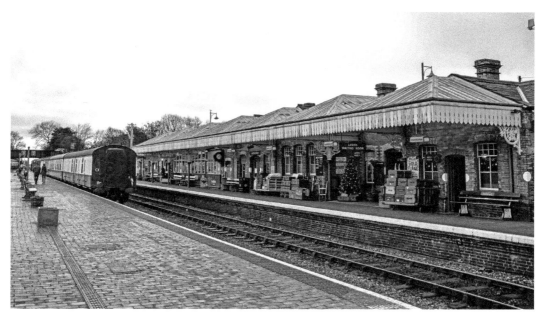

(Nigel Nudds)

Did the growth of tourism all along this part of the coast prompt the arrival of the railways in the town or was it the other way around? A moot point. But what is certain is that there were soon a lot of new houses in and around the station as the town's popularity as a tourist resort grew, along with its excellent reputation for locally caught crab, lobster and whelks.

The station was closed in 1967 after British Rail decided upon a new location for the town to the east of the old site. It wasn't idle for long, however, and in 1975 became the eastern terminus for the North Norfolk Railway – the recreation of a typical post-Beeching branch line, which now operates a service between Sheringham and the town of Holt, which lies inland around eight miles west of Sheringham. The route taken between Sheringham and Holt is known as the 'Poppy Line', christened thus by the journalist and poet Clement Scott who visited the area in Victorian times, referring to it as 'Poppyland' in his column in the *Daily Telegraph*.

The line today is part of the old Melton Constable–Cromer Beach branch line, one that was built with the specific intention of exploiting the growing tourist industry in the area. Riding it today is a sheer delight and you absolutely and positively do not need to be a railway enthusiast to enjoy the gentle run between the two towns. I wasn't when I experienced it for the first time, though I am now, and so be warned – you may be 'turned' as well because of a nostalgic day out along the delightful Poppy Line.

45 Beeston Bump, Beeston Regis
Aka, Beeston Hill.

The village of Beeston Regis is not the place where you would go to if you wanted to see an old Tudor farmhouse or grand Georgian Hall. Its places of historical interest are far more ancient. The tip of that last great ice age glacier previously 'seen' at Salthouse (see No. 38) had a very real say on the physical landscape here as well.

Beeston Bump is a clifftop hill that overlooks the chilly North Sea and village of Beeston. It's a mammoth (this is Norfolk, remember?) 207 feet high and is very much the dominant physical feature on the local landscape. It is, like Salthouse Heath, part of the Cromer Ridge, a remnant of the glacier's northern retreat at the end of the last ice age around 10,000 years ago. Beeston Bump has been subject to the forces of nature ever since with most of its sea-facing side having been eroded away by the end of the 1930s. The constant erosion here has also seen, among other things, the loss of a football pitch and a brickworks. The sea's voracious appetite has, for now, at least been tempered by the introduction of groynes and a new sea wall at its seaward edge.

Much to my delight, Beeston Bump is included on the online site that details all of the hills in Great Britain, meaning that is categorised alongside the rather more emphatic rises at Ben Nevis and Scafell Pike. Yet, spectacular as both are, given the choice of ascents on a June morning, I'd plump for Beeston's Bump every single time.

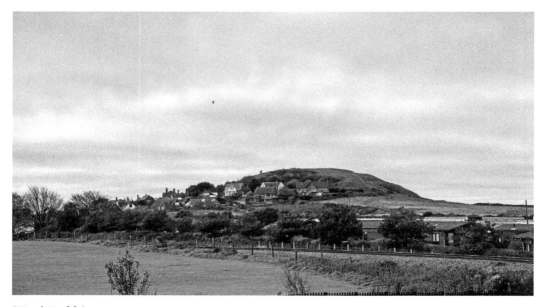

(Nigel Nudds)

46 Pavilion, West Runton
Tribute to a landmark site that no longer exists.

West Runton Pavilion is, alas, no longer with us. Nevertheless, while you cannot visit it, there is at least a blue plaque that celebrates both it and what it stood for at the site where the famous old building once stood.

It was, as far as thousands of music fans were concerned, *the* place to go for live music in the whole of East Anglia – one that saw appearances by some of the most famous and successful rock bands anywhere in the world; a list of the great and the very good that includes AC/DC, Dire Straits, Iron Maiden, Joy Division, Motorhead and the Sex Pistols. Admittedly, some of those names may not mean anything to a lot of you. However, take it from me, they were at the time, very much the A-listers of the music world in the same way that the likes of Adele and, quite unfeasibly, One Direction are today.

And all this appeared in a sleepy North Norfolk village with a population of just over 1,000 people.

West Runton Pavilion was originally a sports and social hall, which, like so many in Norfolk at the time, began hosting dances during the Second World War – events that were very popular among serving US airmen at the time, with the music on offer provided by various military bands.

From the 1960s onwards, the venue became a popular one for the rock bands of the day and, as word spread of its suitability, more and more began to arrange to play one-off gigs there, much to the delight

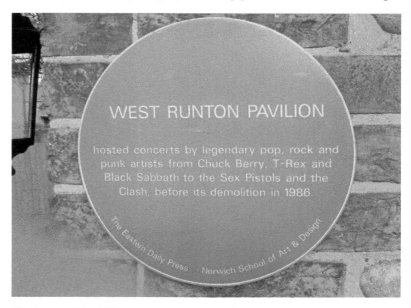

(Graham at The Village Inn, West Runton)

of the local population who, more often than not, were starved of any form of popular mass entertainment. One of the reasons for this had been Norfolk's perceived 'isolation' from the rest of the country. Big name performers originally wouldn't go there for that reason, and yet, ironically, being off the beaten track was one of the attractions to the up-and-coming performers who played there at that time. It meant that they could practise new material and set lists to a lively and enthusiastic audience without having to worry about critical London- or Manchester-based music journalists who, quite simply, couldn't be bothered to make the long journey there to see them.

It also helped that the acoustics at West Runton Pavilion were remarkably good as well!

Sadly, the venue was closed and demolished in 1986. It was, as you will already have guessed, never replaced and all that exists now to commemorate a truly remarkable piece of live music history is a blue plaque on the side of what is now the village pub.

47 Station, West Runton-The Bittern Line
130-year-old station that continues to thrive.

West Runton railway station was opened in 1887 as part of the Midland & Great Northern Railway (not always affectionately known as the 'Muddle and Go Nowhere') line, which linked coastal villages like West Runton to Norwich and, ultimately, all stations to London.

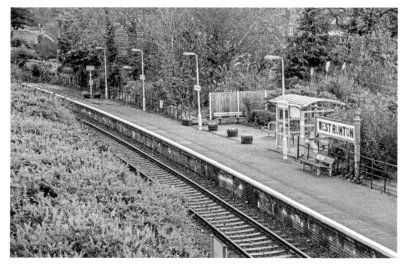

(Nigel Nudds)

It is part of the Norwich–Sheringham line, which is part of the greater national railway network currently owned by Greater Anglia.

The line might, not so very long ago, have been considered a prime candidate for closure, such was the gradual and telling decline of both services and the infrastructure upon which it relied. In order to prevent further loss to services, a recommendation was made to set up a community rail partnership between the local community and the rail network with, as a result, the Norwich and Sheringham Community Rail Partnership becoming known as the Bittern Line in 1997.

The Bittern Line, of which West Runton is a popular 'member' station (passenger usage figures for the period 2014/15 were up on the previous year's total), is a wonderful example of how a transport initiative that has community involvement and input can become a major success. It is one of the most scenic to travel on in England, particularly as it cuts a gentle trail through the Norfolk Broads on its way to the coast from Norwich, then becoming a subtle part of the landscape on the Norfolk coast – there but never intrusive or unwelcome.

For those with a disposition towards all things railway, the line is double track from Norwich to Hoveton and Wroxham station. From there it becomes a single track (except for a passing loop at North Walsham), until it reaches Cromer, from where it passes through West Runton on the way to the end of the line at Sheringham. Happily, the Bittern Line is not resting on its laurels and continues to seek growth and progress. Future plans include the provision of longer trains and either a brand new station at Sheringham, or a connection to the North Norfolk Railway (Poppy Line, see No. 44) there.

You don't need to be going anywhere in particular to take a ride on the Bittern Line. Just call in at one of the stations en route and explore. It's riding the train as it used to be and how it's meant to be.

You can find out more about the wonderful Bittern Line at www.bitternline.com.

Long may it prosper.

48 East Runton
A Runton with a view.

Like so many villages along this beautiful part of the English coast, East Runton was once a small but thriving hub of the local fishing industry. Similarly, like those other villages, it now relies on the year-round tourist trade. It has become one of the most popular holiday destinations in the area, especially for those with a penchant for spending their leisure time under canvas, such is the number of good quality camping and caravanning sites within walking distance of the village.

East Runton's beach is not unlike Holkham's, with great long stretches of sand beneath massive blue skies. The one subtle difference is that, while Holkham, grand as it undoubtedly is, can quite often be rather busy, a walk along the beach at East Runton can extend for many miles without the lucky walker seeing a single soul, let alone a car.

One walk that is locally recommended is the one that takes you down to East Runton village, strolling along it until you come to its companion at West Runton. From there, the route home takes you up and over Incleborough Hill where, from the peak, you can see Beeston Bump sat upon the horizon in splendid undulating isolation.

Just you and the view.

Incleborough Hill, though modest even by Norfolk standards, is owned by the National

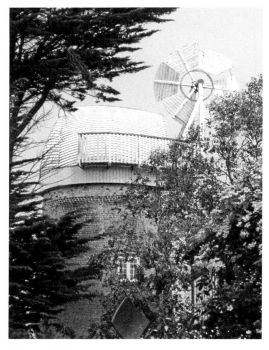

(Victor Gibbons)

Trust and offers a little Highlands mood with the scattering of fell ponies about its well-worn peak. The views are worth the wander, with the coast between Sheringham and Cromer there for you to photograph, paint, or just to admire. There are plenty of views in Norfolk; it's just a case of knowing where to find them.

And this is one of the best.

49 Pier, Cromer
Grade II listed seaside pier.

Records of Cromer's pier date as far back as 1391 although, at that time, the structure in question would have been little more than a plain wooden jetty utilised by the local fisherman. The idea stuck, however, and, as the years and centuries passed, that little jetty evolved into one of the most famous seaside piers in the world.

By 1822, a third pier was built on the site, one that, as it was built entirely of cast iron, would have looked an impressive sight and, more likely than not, been thought of as fairly impregnable, even given the sort of treatment it would have been given by the North Sea. Unfortunately for both the people of Cromer and the pier itself, this

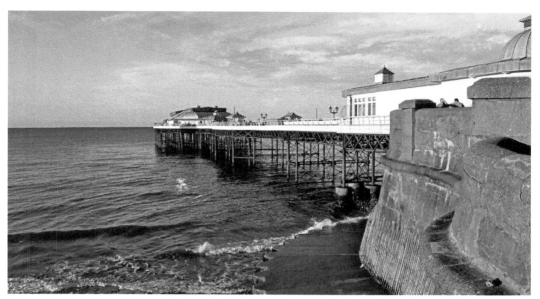

(Simon Moston)

was not the case and, after a typically rumbustious winter's storm in 1846, little to nothing remained of the cast iron monster. Undeterred, a replacement was built, one that, at 240 feet long, was longer than its predecessor. It managed to survive the natural elements with relative ease but was unable to cope with the damage caused to it in 1840 when it was struck by a boat carrying a heavy cargo of coal.

For a while, the good people of Cromer had to make do without a pier. But this was never going to be the case for very long and, by 1902, another new pier was completed and opened to the public. This one was 450 feet long and remains the integral body of the current pier, which is just under 500 feet in length.

The current pier, unsurprisingly, has had its moments of drama, just like those which preceded it.

It has been damaged by storms in 1949, 1953, 1976, 1978 and in 1990 when, during gales, the amusement arcade on the pier was destroyed. Three years after that, an encounter between the pier and a 100-ton rig ended with much of the front half of the pier disappearing into the sea. Cromer, yet again, rose to the challenge of repairing its now very venerable old lady of the sea, which was completed in time for the 1994 summer season and the traditional end-of-the-pier show.

A popular dance tune of 1997, 'Tubthumping' by Chumbawamba, contained the words, 'I get knocked down, but I get up again, you're never going to keep me down' – a song that could be Cromer Pier's theme.

Go take a dry (but breezy) walk out over the chilly brine.

50 Lighthouse, Cromer

Clifftop sentinel that monitors the offshore waters to this day.

Prior to the construction of a lighthouse upon the clifftops at Cromer, passing shipping would have been advised of the proximity and local dangers along this part of the coast via a light that shone at the top of the town's parish church. This was not an unusual arrangement, for, although the light that these would have offered would have been small, they would have been visible for some considerable distance out to sea with, if you recall, a second tower being built at the eastern end of the Church of St Nicolas in Blakeney (see No. 34) for the same reason.

Cromer's original lighthouse succumbed to the ever-encroaching sea and slipped over the edge of the cliffs in 1866. This meant that plans for its eventual replacement, with its loss by the process of coastal erosion having long since been predicted, had to be swiftly acted upon. Fortunately, those responsible for such things were diligent, and the new lighthouse was commissioned and built half a mile away from what was then the existing cliff edge.

The current lighthouse became operational for the first time in 1833, with its distinctive octagonal tower giving it a height of just under sixty feet. It was originally permanently staffed with lighthouse keepers. However, this somewhat romantic and lonely career option ceased in 1990 when the lighthouse became fully automatic, with its operation handed over to a control centre in Harwich, some 100 miles further down the coast.

Old fashioned or not, if I was navigating these shores in a storm, I'd feel a lot safer if I knew there was a person atop Cromer lighthouse looking

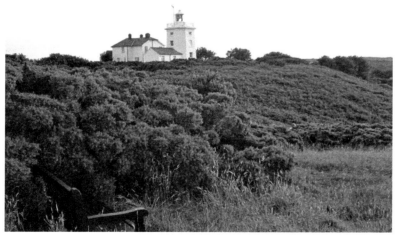

(Simon Moston)

out for me and all my fellow mariners rather than some emotionless bank of flashing lights. But that, as they say, must be progress of sorts. Or, alternately, a money-saving exercise – take your pick.

The lighthouse, being a working building, is not open to the public but, now that you have taken a short detour off the A149 to take a closer look, you will find the area immediately around it fairly accessible.

51 Cromer Hall
The inspiration behind a famous ghostly tale?

Cromer Hall, as majestic and impressive a private residence as you will see anywhere in East Anglia, is not open to the public, so the best the curious explorer can expect of its Gothic facade is a peek across the grounds to the house from Hall Road, which runs south of Cromer and away from our now very old friend, the A149.

I've included it in the book because of its literary connections, which have its origins in a visit that was made to the house by the great Sir Arthur Conan Doyle, famous for his stories about the detective Sherlock Holmes and his redoubtable companion, Dr Watson. Doyle had returned from South Africa in 1901 suffering from typhoid fever. In an attempt to hasten his convalescence, he took a golfing holiday in Norfolk, staying at the Royal Links Hotel in Cromer. There seems little doubt that Doyle, hungry for stories and local folklore wherever he travelled, would in time have heard about 'Black Shuck', the fearsome spectre of a large dog that patrols the lonely beaches and marshes of North Norfolk, its gaze reputed to bring death to whoever looked upon it.

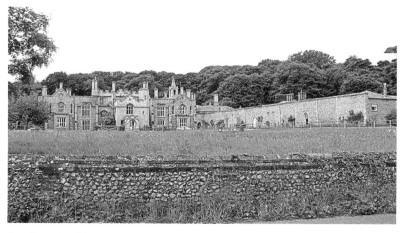

(Evelyn Simak)

The legend of Shuck certainly seemed to have an impact upon the writer who, soon afterwards, presented perhaps his most famous story, *The Hound of the Baskervilles,* to his eager public. The description he gives of Baskerville Hall, the fictitious house in Devon where the story takes place, very closely matches the appearance of Cromer Hall; so it would seem that Doyle spent part of his stay in Norfolk getting the details of the story together and, having seen Cromer Hall, decided that the Gothic style in which it was built suited the house of the tortured Baskerville family very well.

Among the building's striking features are its tall, octagonal stone chimneys and matching octagonal tower on its east side. It was built in 1829 by the architect William Donthorne, the man who was also, if you remember, the man responsible for the Corinthian column in the grounds of Holkham Hall (see No. 27).

52 Lifeboat Station, Cromer
King of Piers.

Cromer has long been famous for its lifeboats and the men who have served upon them, not least the great Henry Blogg (1876–1954), the most celebrated and decorated man in lifeboat history.

The main RNLI station at Cromer is situated at the head of the town's iconic pier. It is a modern facility, having been rebuilt in between 1997 and 1999. The town's previous station, which dated from 1923, was

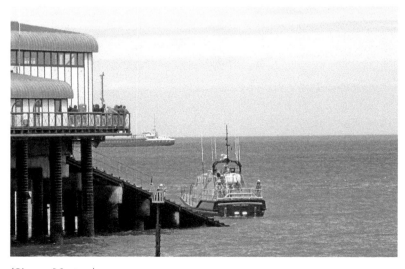

(Simon Moston)

removed and relocated in the Suffolk town of Southwold, where it is used as a museum. The actual lifeboat service at Cromer, however, goes back further than the date of that building's initial construction – indeed, it predates the RNLI itself, which was formed in 1824. It is Blogg, however, along with another local man, Henry 'Shrimp' Davies, who are the two main names that will forever be associated with the lifeboat service at Cromer as well as the history of the RNLI.

Blogg's bravery and dedication to the cause is best illustrated in how his service has been recognised by the RNLI. He was awarded the institution's gold medal three times and its silver medal four times, as well as the George Cross and British Empire Medal.

The station you will see at Cromer today cost approximately £3 million to build and set up, with the money needed coming from private donations and bequests. Its most recent major refurbishment was in 2015 when the original pitch pine, which was used as the outer cladding for the new building, was removed and replaced with Siberian larch, a very durable and resistant timber that should be able to resist the very worst of the North Sea elements for at least sixty years.

53 Railway Station, Cromer
Three became two became one.

As befitting a significant fishing port and holiday resort, the town of Cromer has had three railway stations since the service first arrived in the town in 1887, with the first of these, Cromer Beach (no prizes for

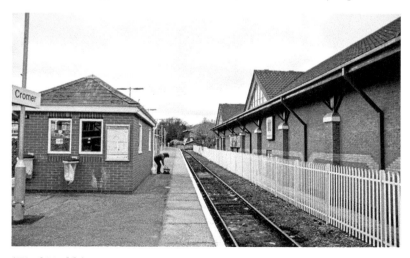

(Nigel Nudds)

guessing what important part of the town that might be near) being renamed Cromer after it outlived the other two stations, which have since closed.

It's based on the Holt Road and is a distinctly modest affair that belies its status as an important terminus for tourists throughout the year, whether they are visiting for the day or longer. Station user numbers have, sadly, declined over recent years, from 194,000 in 2012/13 to 187,000 in 2013/14 and 183,000 in 2014/15.

The decision to focus all of Cromer's resources into just the one station came about as the town's popularity as a holiday resort declined in the years after the end of the Second World War. The logical decision would have seemed to have been to retain the station known as Cromer High, which, with its excellent facilities, would easily have served as the town's single railway hub. It was, however, in the minds of those who matter, a little too far away from the town centre and all that it had to offer the visitor – a beach, fish and chip shops, amusement arcades, hotels and bed and breakfasts aplenty. So, the decision was made to retain Cromer Beach and to close Cromer High, which duly saw off its last paying passenger in 1954.

The third station in the town was Cromer Links Halt, which opened in 1923. It was a basic facility that contained little more than a single wooden platform, a couple of wooden benches and some oil lamps. Its main purpose was to serve golfers at the nearby Royal Cromer Golf Club (links golfers are used to hardship given the weather conditions they often have to play in, so this was probably not a problem for them) but, with the station providing no direct link to Norwich, it became something of an indulgence so far as British Railways were concerned and was closed in 1953.

The current station is part of the Bittern Line, with trains heading westwards from Cromer having just two more stops before it terminates at Sheringham, the preceding stop to that being West Runton. It's a rather modest affair, which is a shame as it seems that a noble town like Cromer should have a splendid station, one that reflects the great canvas of the surrounding countryside.

54 Felbrigg Hall

Seventeenth-century country house noted for its Jacobean architecture.

The Felbrigg estate used to belong to the family of the same name but, upon the death of the last Felbrigg in the fifteenth century, it passed to the Wyndham family, including a John Wyndham (1558–1645) who,

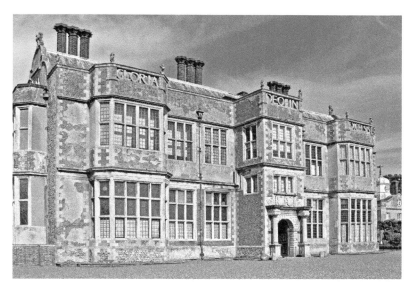

(Mark Oakden at www.tournorfolk.co.uk)

along with his son Thomas, was largely responsible for the fine house that you can visit to this day. It is particularly noted for its Jacobean architecture and fine Georgian interior.

John Wyndham is quite an important historical character, having more than played his part in defending the nation from the threat of Spanish invasion through helping to organise and lead a massed defence force that was based in the West Country in 1580, consisting of 12,000 ready and very willing foot soldiers. Wyndham also served as a Justice of the Peace in his native Somerset, and so all in all was very much the 'cock of the walk' in his home county, an influential and relatively prosperous man who had a large circle of friends.

It must, you would assume, have been quite a wrench for Wyndham to sacrifice all of that and decamp to Norfolk when, in 1599, he inherited the Felbrigg estate from his father's cousin, Thomas Wyndham. He clearly saw it as his duty and swiftly settled down to everyday life at Felbrigg, establishing in the process, along with his wife Joanne, a dynasty that, given they had fourteen children, could well have lasted to the present day. This, however, was not to be the case and, with the death of Robert Wyndham Ketton-Cremer in 1969, the house was passed over, as per his wishes, to the National Trust.

The current Felbrigg estate covers an area of 1,760 acres of parkland that includes the 520-acre Great Wood. Its gardens include a traditional orangery, formal lawns and a walled garden. The gardens are also home to the national collection of colchicums – a flowering genus that includes the crocus family.

55 Southrepps
Ancient village torn in two by the Black Death.

The quiet (but aren't they all?) village of Southrepps is split into two halves, namely Upper Southrepps and Lower Southrepps, renamed locally as Upper Street and Lower Street. These two halves are separated by open farmland, a situation that is believed to have come about because of the Black Death, the deadly plague that laid waste to much of the country in the fourteenth century, splitting, in its wake (and quite literally in this case) communities asunder in the process.

The farmland that divides the village is highly prized, with both a favourable climate (Southrepps enjoys a low annual rainfall and, at ten degrees centigrade, relatively high mean annual temperature) and geology in its favour. Its combination of glacial sands and gravels in combination makes for very well drained and fertile soil. The village has always been a highly productive one in terms of crop yields, especially in the decades following the end of the Second World War, when crop production and output, both in and around the village and along this stretch of coast, became more and more intensive.

The village also contains a Site of Specific Scientific Interest (SSSI) as well as a nature reserve. Southrepps Common covers an area of around thirty acres and supports a varied selection of wildlife habitats, including woodlands and reed beds. It is also accessible to the visiting public, with a boardwalk allowing entry to the most important parts of the site to all, including wheelchair users. There are known to be at least 160 different plant species across the site, including a selection of orchids.

Southrepps is a very community-minded village that isn't, unlike some, content to sit back and let the world pass by unnoticed. There

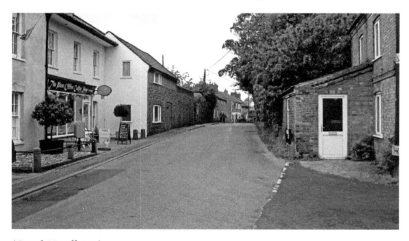

(Carol Needham)

is, for example, a Southrepps Society whose 2016 programme included a barbecue, guided walks in and around the village, and an intriguing talk titled 'Managing Pests and Diseases the Natural Way'. The village also hosts a Classical Music Festival every year with the objective of bringing 'leading young artists to make music of the highest order in a beautiful corner of rural Norfolk'.

Beautiful. And unspoilt. Let us all hope it remains that way. Treat Southrepps with care – it is one of a vanishing kind.

56 North Walsham & Dilham Canal
Or how to get cabbages from A to B.

Even though Norfolk is richly illustrated and characterised by water, canals are not a feature that you would necessarily associate with the county, instead perhaps more readily identifying them with parts of the Midlands or even London. This is a county that contains not only a high number of rivers but has, in the Norfolk Broads, 120 miles of navigable waterways and sixty-three broads. Thirteen of these are open to navigation, which would logically suggest there must be numerous canals linking one of those broads to a nearly river or adjacent broad.

It may well surprise you, therefore, to learn that, despite all that water sloshing about the county, there is only one artificial and locked sailing canal in Norfolk. And this is it.

The North Walsham & Dilham (NW&D) canal was authorised in Parliament in 1812. Work did not start on its construction until thirteen years later. Unlucky for some? Maybe. It was never a commercial

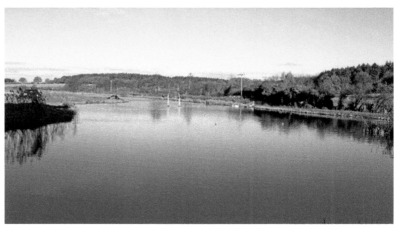

(Paul King)

success and was frequently sold to a new owner who would have, no doubt, grand plans for its future, only for them to become as disillusioned with the canal as their predecessor and to sell it on again. This tapestry of owners would have included many of the millers who owned the watermills spread out along the canal's length (six in total), including the two bone mills at Antingham, where it commenced its near nine-mile run to a junction with the River Ant at Smallburgh.

With the advent of swifter means of freight, sounding the death knell to canals in terms of their economic viability, NW&D's active 'life' as a commercial venture ceased in 1934. It was, however, readily recognised as being a possible site for leisure activity; the combination of this with its value as a historical site eventually lead to major restoration work being carried out along its route from 2000, when the East Anglian Waterways Association (EAWA) started to recruit volunteers with an interest in preserving the canal. Eight years later, the newly formed North Walsham & Dilham Canal Trust started to run joint working parties on the canal with EAWA. In 2009, part of the canal was sold to the Old Canal Company, who have assisted with the restoration of two locks as well as the ponds that laid in between them.

NW&D was primarily built for the transport of coal. It proved, however, to be cheaper for coal to be transported overland, which meant that the canal had to find other cargoes, i.e. those that went to and from those six watermills that lined its banks, one of which was a wherry that carried cabbages to Great Yarmouth.

57 St Nicholas Church, North Walsham
Tower tumbled by over-enthusiastic campanologists.

The Parish Church of St Nicholas in North Walsham is perhaps more well-known for the part of it that isn't standing, rather than for any part of the rather splendid church that exists today in its entirety.

Its tower was once a very prominent local landmark that, topped with a steeple, reached a height of 180 feet; in comparison, the celebrated steeple at St Mary's in Snettisham (see No. 8) stands at a 'mere' 172 feet. One, maybe apocryphal, story tells of how the good folk of North Walsham, incensed at the completion of the 160-foot-high tower at nearby Cromer's Parish Church, vowed to steal a little of that town's thunder by declaring that, 'Our church is bigger than your church', and setting out to prove exactly that.

An act, it would seem, of some folly. The tower collapsed in May 1724, on the morning after the bells had been rung for most of the

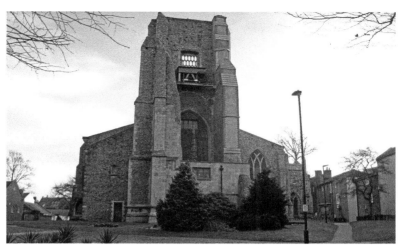

(Paul King)

day to mark Ascension. God fearing or not, the resultant noise and vibration did more to harm the structure that the bells resided in, than remind everyone in the town what day it was.

Despite this lasting damage, escalated in 1836 when the north side of the existing ruin collapsed during heavy gales, the building today remains one of the largest parish churches in the country and demands a visit – but not only for the opportunity to gaze upon the ruins of that most ambitious of towers but also to see several unusual features within the church itself, including an extremely ornate marble sarcophagus of Sir William Paston.

Paston, named after the Norfolk village where he was born, which lies around four miles north-east of North Walsham, worked as a lawyer, investing, as he did, in several properties and accumulating no little wealth in the process. He is regarded as the founder of the Paston family's fortunes, which include the collection of letters, documents and state papers exchanged between members of that family from 1422 to 1509, and collectively known as the Paston Letters.

58 Peasants' Revolt, North Walsham
The bishop who preferred war to peace.

There is, of course, no daily recreation of the Peasants' Revolt in and around the bustling streets of North Walsham, and neither does the town recreate the event's place in its history with a series of attractions and gift shops that are prefixed with the phrase 'Ye Olde' in a sad attempt to push whatever tat is being sold into the hands of the gullible.

North Walsham is above all that nonsense. As are you.

So how about considering a visit to North Walsham's very own field of combat? It isn't, admittedly, the Battle of Naseby, but that doesn't mean the encounters that have happened here are any less important a part of English history.

The Battle of North Walsham (one Hollywood hasn't won for the US yet) was fought near the outskirts of the town in June 1381. It saw a large, unruly and rebellious group of peasants take on the forces of one Henry le Despenser, the then Bishop of Norwich. This little-known battle is historically significant as it was the last major resistance of any kind during the more widely documented Peasants' Revolt, which took place throughout England during that year.

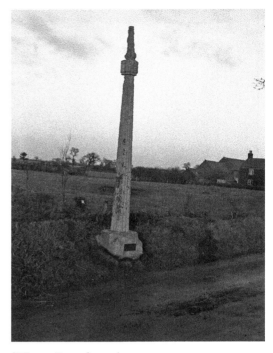

(Wayne Beauchamp)

Despenser might have been a bishop, but he did like a good scrap. Before being consecrated, he had enjoyed a happy campaign in Italy in 1370 before leading a crusade to Belgium in 1383, one that was committed to defend English political and economic interests in the area.

An army of poorly organised and armed peasants was therefore highly unlikely to cause le Dispenser too many sleepless nights. And so it proved. His original fighting force was made up of men who were loyal to him; however, assured that they would be booking their ticket to heaven if they sided with such a man of God, he soon found more and more men joining him on this rather more provincial affair, many of whom would, quite likely, have been friends and comrades of those they would be fighting against.

The power and allure of the Church was a powerful thing, though, and so much more so if it was led by a charismatic man.

That final battle at North Walsham saw le Dispenser's army thoroughly rout what was left of the peasant rebels, led by Geoffrey Litster, the self-titled 'King of the Commons' who was captured and executed at the end of the battle.

North Walsham commemorates the battle with three medieval stone crosses, one of which now doubles up as a parish boundary marker. It is situated off the Norwich Road on Toff's Loke (a loke is a short and narrow lane that often comes to a dead end).

59 Paston Grammar School, North Walsham

De mieux en mieux pour tout – 'From good to better everywhere'.

The original Paston Grammar School was founded in 1606 by the prominent magistrate, landowner and libertarian Sir William Paston. He was a well-connected man who was also a member of an important local family (see No. 57) and a man destined to end up immortalised in marble at the local church – no mean feat at a time when local merchants preferred their wealth to be spent on the church itself rather than on indulgent internal decor.

That school saw 400 boys in attendance, many of whom would progress to Gonville & Caius College in Cambridge, itself then nearly 300 years old and considered a centre of excellence for Paston alumni who wanted to follow a career in medicine. Upon the arrival of the railway at North Walsham in 1874, the catchment area for the school grew, as did its reputation further afield, with its companion school, the North Walsham High School for Girls, opening its doors at the beginning of the twentieth century. Thus, humble but forever pleasant North Walsham became an educational centre of excellence to rival Norwich.

Paston Grammar School for Boys could not shine as a wholly independent entity forever, however. Sweeping education acts in 1902 and 1944 saw the school become, as a consequence, one that was managed by the local authority, ultimately as a voluntary aided (a state-funded school where a foundation or trust, most usually one relating to a church, has a substantial influence in the school's day-to-day management) grammar school for boys.

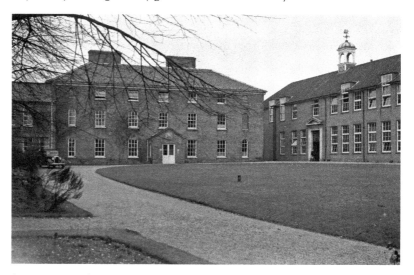

(Wayne Beauchamp)

Its remit and heritage had, therefore, not changed or been swept aside. But its roots most certainly had.

The two schools eventually merged in 1984, becoming, in the process, the present-day sixth form college, which caters for boys and girls from the ages of sixteen to eighteen, with undergraduate life remaining a popular option for many, just as it did over 700 years earlier.

One of the school's most famous ex-pupils is Horatio Nelson, who attended it from 1768 to 1771 before joining the Royal Navy. There, it must be said, he made quite a name for himself as the one-time sickly and rather weedy son of a North Norfolk Rector – one whose familiarity with the rigours of life would have been first met at Paston.

60 Worstead
Flanders comes to Norfolk.

Worstead is a village that has been there and done that. More to the point, it has done more than worn the t-shirt. For it very probably would have helped to make it as well.

It's an ancient settlement, one that was known in the *Domesday Book* as 'Wrdesteda'. It had, prior to the compilation of that great manuscript, been presented to the Abbots of St Benet's Abbey as a gift by King Canute, he who has forever been inaccurately portrayed as the man who tried to stop the tide from coming in. The village, even then, clearly had a lot going for it as, around 300 years after Canute's act of giveaway flamboyance, a host of weavers, fresh out of Flanders and looking for a loom with a view, saw the village as a pleasant enough one to settle in, having been encouraged to make England

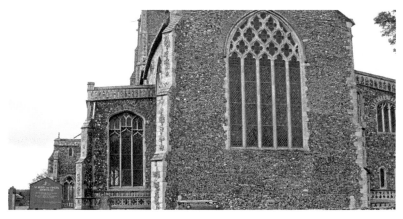

(Paul King)

their home by King Edward III, who had recently married a Flemish princess, Philippa of Hainault.

Happy in their new home, the weavers soon got down to work with Worsted cloth the end result.

Worstead could also claim to have partially contributed to the origins of the noun 'stuff', which is how the original cloth, lightweight and with a coarse texture, was known. It is, therefore, to distinguish between worsted cloth and that derived from wool that the natural crimp (it compresses into small folds or ridges) of wool is removed in the process of spinning a worsted yarn. This means that the straight fabric can be tightly compressed and is therefore ideal for manufactured surfaces that endure a great deal of wear – for example, carpets, gloves and baize.

Sadly, although worsted cloth is now internationally renowned and linked to the village where it originated, it is no longer manufactured there, with the last weaver from the village, John Cubitt, dying in 1882. Weaving and spinning demonstrations are, however, part of the village's hugely popular annual festival, which is held at the end of July.

61 Dilham
Castle, Church & Dill.

Dilham sits on the River Ant, slightly astray from the A149, and earns its mention here for the reason behind its name: dill, the popular culinary herb used along with chives and parsley that was once commercially grown here in substantial quantities.

Dilham is also, of course, one part of the North Walsham & Dilham Canal (see No. 56) as well as acting as the limit of navigation on the Norfolk Broads for larger boats. Happily, smaller and less intrusive craft can gladly take to the aforementioned canal all the way to Honing.

I've already mentioned, in Titchwell (see No. 16) and Burnham Deepdale (see No. 22), two of Norfolk's 124 remaining round tower churches and, in Dilham, I happily mention a third: namely Dilham's St Nicholas, one of the many in this part of the county that have been named for him. This is not surprising really, as he is the patron saint of sailors – hence the proliferation of churches in fishing villages across the country that are named in his honour. As far as 'his' church in Dilham is concerned, St Nicholas might well have been concerned to note that it has, to say the least, had an eventful history. The signs of what was to come first came to light in the eighteenth century when cracks on the tower led to its eventual demolition. However, this new build and any and all additional patching-up that was also done at the

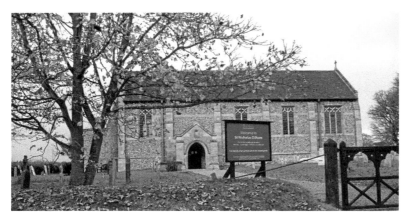

(Paul King)

time proved to be inadequate. The tower, originally restored in 1775, was pulled down eighty years later and replaced with a new round tower, which only lasted for a further seventy years.

Another extensive rebuilding programme commenced in the 1930s, one that, for reasons of cost and practicality, eschewed perceived fripperies altogether. It's built to a simple and modernistic design that is, for all that, pleasing on the eye and as worthy a place of worship as any cathedral – even if the tower that accompanies the main building is truncated and barely reaches halfway up the existing west wall of the church.

The remains of Dilham Castle can be seen in the grounds of a farm in the village.

⑥ Barton Broad, Barton Turf
An environmental success story.

Barton Broad, the second largest of the Norfolk Broads (Hickling Broad is the largest), was dug and utilised as a source of peat during the Middle Ages. Once the supply had been exhausted, the nearby River Ant was diverted into and through the pit, allowing the 'new' broad, which is very shallow in places, to be navigated by small river craft. Barton Broad was, up until the middle of the last century, renowned for its clear water and the rich mix of animal and plant life that came with it.

Over time, however, the broad gradually became silted up with sediments that were rich in nitrate and phosphate – bad for the quality of the water and the welfare of its wildlife, with fish numbers declining in particular. This was halted by the Broads Authority in the 1990s, who

78

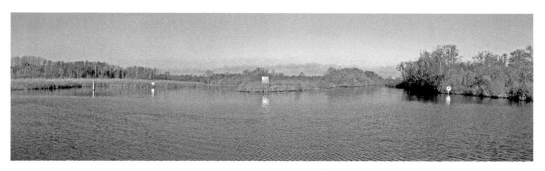

(Mark Oakden at www.tournorfolk.co.uk)

facilitated a dredging programme in order to remove the sediments and, with it, improve the quality of the water. This programme was accompanied by works carried out to the nearby sewage treatment works further up the River Ant. The project was deemed to have been a success, with water quality and depth substantially improved as a result. The whole process took four years of pumping and removed an estimated 50 tons of phosphorous from the broad.

Barton Broad is in the care of the Norfolk Wildlife Trust as a nature reserve. It is reputed to be one of the places where Horatio Nelson learnt to sail (Nelson is also said to have learnt to sail in the narrow creeks in and around Brancaster Staithe Harbour – see No. 21).

63 St Michael & All Angels Church, Barton Turf
Prepare to be astounded.

St Michael & All Angels Church in Barton Turf, like so many of Norfolk's fine old churches, dates back to the fourteenth century. And, while the building itself – tree framed and less than a mile south-west of the village – is an attractive one, it is what lies inside that is of great local, national and, indeed, international interest as far as ancient church decor is concerned.

St Michael & All Angels' 'secret' (for, like the magnificent St Margaret's Church in Cley, if it were a church of Florence or Rome, it would be an object of international renown) is its twelve-panelled rood screen. These are typically carved from stone or, in this case, wood, and are designed to mark the border between a church nave and its chancel. Barton Turf's dates from around the middle of the fifteenth century and appears to have been influenced by previously existing work that can be found in Flanders. With that Flanders influence, it is interesting to note that the screen dates from around the same time that the

79

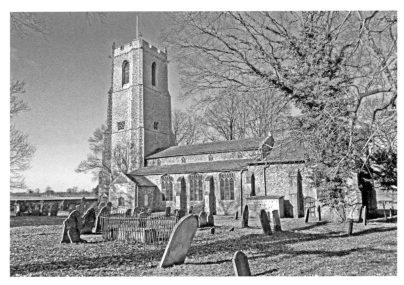

(Mark Oakden at www.tournorfolk.co.uk)

weavers from Flanders were arriving in Norfolk and settling in and around the Worstead area (see No. 60). Among the depictions on the screen are those of three saints, namely Apollonia, Zita and Barbara, as well as some images that were defaced as papist symbols during the English Civil War. While Saint Apollonia is not the most well-known of saints (among those who regard her as their patron are dentists), she does seem to have rather a large following in Norfolk, for, as well as in Barton Turf, there are images of her at churches in Docking, Horsham St Faith, Ludham, Norwich (two) and Sandringham.

The twelve-panel screen sits across the chancel arch, while a slightly less ornate four-panel screen is positioned across the south aisle chapel – not that you, or anyone, need directions on where to locate them once you are in the church.

It's a work of art that deserves – no, demands – to be seen. And it sits, in relative obscurity, in an otherwise modest Norfolk church.

Sometimes we don't know how fortunate we are.

64 Wayford Bridge, Stalham-River Ant
A river that will be familiar to thousands of school children.

Wayford Bridge gives the curious traveller an opportunity to cast their eyes upon the River Ant – for me, the most romantic and special of all of Norfolk's rivers, for reasons I will explain further down the page.

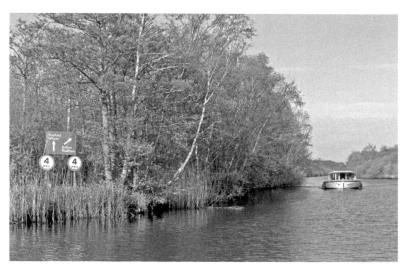

(Mark Oakden at www.tournorfolk.co.uk)

Height restrictions under the bridge mean that it is the most northerly point of navigation on the Norfolk Broads – at least for boats that are over 7 feet 6 inches above the water line. Smaller broads boats are able to pass under the bridge and continue to Dilham, albeit in a channel that is narrow and offers limited opportunities to turn around – a route for only the most experienced or confident of leisure sailors.

For me and countless others who were fortunate enough to have stayed there during their schooldays, the River Ant will forever be synonymous with How Hill House, the Norfolk Broads study centre that is situated within the grounds of the Broads National Park and runs, among other things, residential environmental courses for groups of children, with walks and boat trips along the River Ant very much part of the daily activities. The house, the surroundings and the River Ant itself are a delight and, if your first glimpse of the river from the Wayford Bridge on our old friend the A149 tempts you to want to take a closer look, How Hill would be as good a place as any to start your exploration of a magical waterway.

65 Stalham
The Museum of the Broads.

The market town of Stalham is home to the Museum of the Broads, which was moved there and opened in 2000 with the aim of bringing 'the history of the Broads alive for *(both)* locals and visitors to Norfolk'.

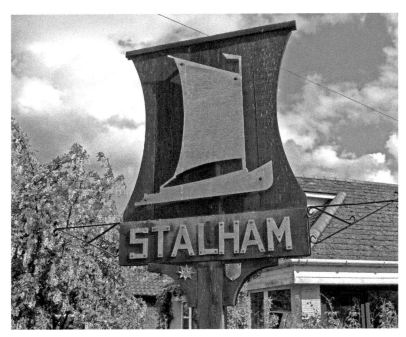

(Mark Oakden at www.tournorfolk.co.uk)

The museum's extensive collection includes objects that illustrate how people have exploited both the surrounding land and waterways for food, transport and, ever more significantly, leisure. Many items, including some of the museum's boats and the tools used for their construction, are unique, as are many of the archive photographs they possess – treasured items that give a special insight into the lives led on the Norfolk Broads. The collection includes a display of the different types of toilet that are used on the boats that sail upon the Broads, an important detail when you consider the environmental implications of all that potentially harmful material if it isn't managed properly!

The museum has four buildings: namely the Barton, Surlingham and Wroxham Buildings, and the Oulton Boat Shed. They also possess a steam launch, *Falcon*, which is moored on their quayside.

Stalham is ideally situated as one of the major starting points for Norfolk Broads boating holidays. It is also, however, just a few miles from the nearest sandy beach, and so can offer the explorer the very best of both worlds. The River Ant at Stalham flows downstream towards other points of interest, including How Hill (see No. 64), Barton Broad (see No. 62) and Ludham Bridge.

The town itself hosts a small market every Tuesday as well as a farmers' market on alternate Saturdays.

66 Hunsett Mill, Stalham
The most photographed structure on the Norfolk Broads?

One of the many landmarks scattered along the length of the River Ant is Hunsett Mill, which sits on its east bank around one mile north of Barton Broad (see No. 62) and a little over the same distance south-west of Stalham (see No. 65). The windmill, like so many of the remarkably well-preserved buildings of this type throughout Norfolk, is a Grade II listed building.

The windmill was built in 1860 and stands just under 40 feet high. Its outer appearance has been extremely well preserved and it retains much of the look and character it would have had at the time it was built, even though work that was performed on its interior in the 1960s removed any evidence that it ever was once a working mill. Unfortunately, at least as far as any of the resident mill keepers were concerned, little due care and consideration seems to have been given to the accompanying house, as it has needed to be rebuilt and modernised on several occasions, most recently in 2008. That most recent redevelopment did, however, pay rich dividends when it was shortlisted for the Structural Awards of 2009 as one of the projects with the most innovative structure in the category that dealt with residential structures – high praise indeed, and well deserved.

Hunsett Mill is not the easiest of buildings to reach, as it is situated right at the most northerly edge of the Norfolk Broads network and has no public road access. In addition to that, it is privately owned and so, sadly for you and me (but most fortuitously to the lucky owners) the best view you will get of it will either be from a distance or via one of the myriad postcards featuring the sights of the Broads – Hunsett

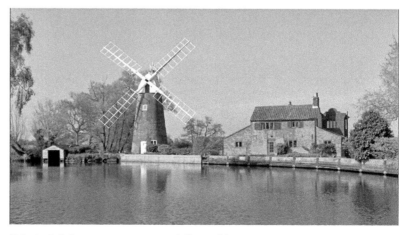

(Mark Oakden at www.tournorfolk.co.uk)

Mill being among the most featured and popular subjects for any photographer in the area.

Hunsett today would not have been the first mill on the site. The present windmill includes a stone within its structure that is marked, HUNSETT 1698, this presumably referring to the date of construction of the original windmill.

⑥⑦ Sutton Mill, Sutton
A giant among windmills.

The (then) eight-floored Sutton Windmill, built in 1789, has the distinction of being one of the tallest surviving windmills in Britain as well as being one of the most striking and physically impressive landmarks in the whole of Norfolk.

It was used for producing animal feed until 1940 when, sadly, it began to fall into a state of disrepair. For three decades, it was the base of operations for the Sutton Windmill and Broads Museum; its cavernous interior space perfect for the collection of bygone artefacts that were displayed there. It was then sold to a historical attractions company in 2006 but they only retained an interest in it for two years before selling it again. Four years later, many of the museum's remaining artefacts were sold at an auction and the fate of this wonderful building was left very much in the balance.

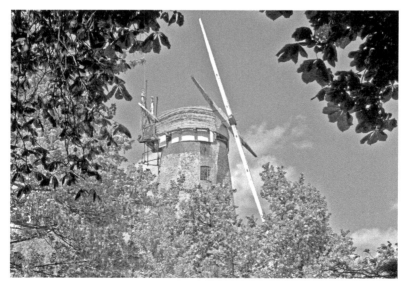

(Mark Oakden at www.tournorfolk.co.uk)

The rather bleak outlook was not made any easier when, in December 2013, the windmill was badly damaged in a severe winter storm, one that was so violent that it managed to tear large chunks of timber off the windmill's cap. Restoration on this wonderful mill, which is privately owned, is now ongoing.

Sutton Mill was not without the sort of incident that might have seen it lost forever during its working pomp either. It suffered a serious fire in 1861 (said fire led to the original eight-storey mill having an extra floor added, bringing it up to the nine-floored behemoth it remains to this day) and was hit by lightning in 1875 and 1940, the latter strike causing a fire that ultimately led to the end of the mill's working life.

68 Bridge, Potter Heigham
A bridge with attitude.

Inexperienced sailors beware. The medieval bridge at Potter Heigham, a village built on the River Thurne, is not only one of the most well-known features on the Norfolk Broads but is additionally notorious for being the most difficult to navigate. It dates from 1385 and consists of three arches, two of which are triangular and date from the fourteenth century, while the other, central arch, is circular and built at a time when the bridge underwent some modernisation in the fifteenth century. It is, hardly surprisingly, a Grade II listed structure as well as a scheduled ancient monument. To preserve it from the relentless rush of traffic, especially at the peak of the holiday season, a modern bridge, the A149 Potter Heigham Bypass bridge, was constructed. This is also a little bit easier to navigate under, providing headroom of 7 feet 9 inches at its highest point. The headroom at the centre of the old bridge is only 6 feet 8 inches.

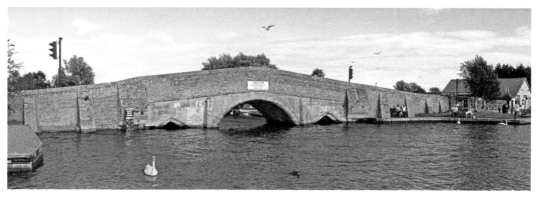

(Mark Oakden at www.tournorfolk.co.uk)

Such is the challenge of passing under the ancient bridge (navigable for only small cruisers at low water) that help is on hand for the unwary at an adjacent boatyard with a fee payable for that expert assistance and local knowledge.

Holiday trade aside, Potter Heigham may seem a relatively quiet and insignificant village, but it was far from being regarded as that during both world wars. Part of Hickling Broad, which lies within the village parish, was used as a seaplane base from 1916 to 1918, while part of the Ludham Airstrip, part of a large Second World War airbase, was also situated within the parish. In addition to that, there were also several defensive structures located around the old bridge – as if crossing under it wasn't difficult enough in the first place!

69 Church of St Nicholas, Potter Heigham
Exquisite.

The church at Potter Heigham is one of the many along this part of the Norfolk coast that are dedicated to St Nicholas, the patron saint of fishermen and children. Its earliest surviving feature is its round tower, which dates to the twelfth century, while the octagonal extension that was later added to the tower was added in the fourteenth century.

Bearing in mind the 'fate' of some Norfolk churches after they'd put into place plans to make their building stand out from all the others (see what

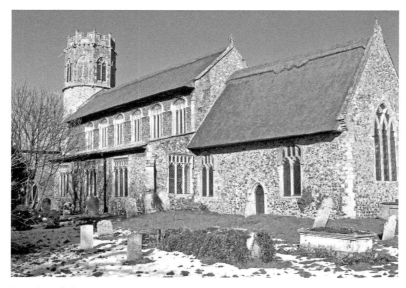

(Mark Oakden at www.tournorfolk.co.uk)

happened at St Nicholas Church in North Walsham, for example), the fact that this structure not only survived but is in a relatively good state of repair says a lot for the planning and craftsmanship that went into that extension at the time. It might also say a lot about the durability of round tower churches, the example here being one of 124 that remain in the county.

But don't dally outside this rather lovely old church – look inside as well.

St Nicholas at Potter Heigham has a fine hammer beam roof and the only brick-built font in Norfolk. Then there are the famous *Seven Works of Mercy*, a set of wall paintings that can be seen on the south aisle. The church is a Grade I listed building ('a building of exceptional interest') that was awarded this well-deserved status back in 1955. Both on the outside and inside, it is an absolute jewel that deserves any and all accolades that it gets.

70 Repps with Bastwick
Calm within the boating melee.

Repps with Bastwick comprises of two adjacent villages, Repps and Bastwick, that are situated on the A149 around 10 miles north-west of the road's denouement at Great Yarmouth, a civil parish that borders the River Thurne with Bastwick at the south end of the bridge, which carries the A149 over that river.

The village is linked with Potter Heigham and Thurne by the Weavers Way, a fifty-six-mile-long footpath that runs from Cromer to

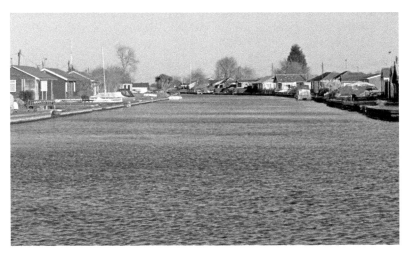

(Mark Oakden at www.tournorfolk.co.uk)

Great Yarmouth, also passing through, among others, North Walsham, Worstead, Stalham and Potter Heigham. It was the A149 of its day in many ways, a well-travelled route that made access to many of these important Norfolk towns and villages relatively safe and direct for the busy wayfarer. Its name comes from the cloth industry that was, as we have already discussed (see Worstead, No. 60), so important to the region before the advent of the industrial revolution.

The round-towered village church of St Peter (in Repps), one of those doughty 124 survivors of its type in Norfolk, is a striking one that merits further investigation. Parts of the tower date back to the thirteenth century; however, like so many churches, its interior was subject to over-enthusiastic restoration towards the end of the nineteenth century although, thankfully, St Peters survived some of the excesses that were inflicted upon other churches in the early part of the twentieth century.

In a part of the county that occasionally looks as if it will drown under the weight of holiday fervour, Repps with Bastwick comes as a welcome oasis of calm, unassuming villages in pretty countryside that have, as yet, refused to yield to some of the extravagances that are very evident within their neighbours.

71 Rollesby Broad, Rollesby
Gone Fishing.

The village of Rollesby sits adjacent to Rollesby and Ormesby Broads, with the A149 crossing the former as it heads out of the village and eastward, towards the neighbouring village of Ormesby St Michael.

Rollesby Broad is an integral part of the larger collective of waterways known as the Trinity Broads, which, in addition to Rollesby, also include Ormesby and Filby Broads as well as two smaller broads named Ormesby Little Broad and Lily Broad. The Trinity Broads are on a tributary of the River Bure, with Rollesby, Ormesby and Filby Broads connected to the main network by the Muck Fleet.

Confusing? It can be. However, what it all illustrates is how glorious a patchwork of waterways the Broads are – myriad rivers, tributaries, broads and inlets, all linking up, like an aquatic network of roots. And all because of the actions of peat digging and the resultant flooding of those workings. The Broads, as well as some surrounding land, were constituted a special area with a level of protection equivalent to that of a national park by the Norfolk and Suffolk Broads Act (1988), with the Broads Authority, the statutory authority responsible for managing the area, becoming operational a year later.

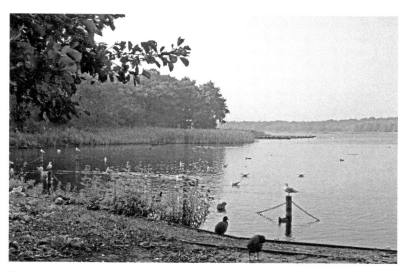

(Mark Oakden at www.tournorfolk.co.uk)

One great delight/hindrance (opinion will be divided) of much of the Trinity Broads network is that they are not connected by any true navigable link to the main river system and are therefore very much undisturbed by the trappings of commercial and leisure activity. Being 'land locked' in this way means that, other than the occasional canoe or rowing boat, they are free from sailing craft, which means that the most popular activity on Rollesby Broad is fishing, with several boardwalks created to allow easy access for anglers.

72 Lifeboat Station, Caister-on-Sea
Where lives are saved.

Caister-on-Sea lifeboat station is one of those buildings (unlike some in this book!) that cannot easily be missed, as it is brightly coloured and cheerful in much the same way as the large barns you can see dotted about the grain belt in the heartlands of the United States.

Caister's lifeboat service is one of the few remaining in the UK that is independent from the RNLI. It was first put into service in 1791 and, although it came under the governance of the RNLI from 1856 to 1969, it is now a registered charity that operates both inshore and offshore lifeboats. The service currently has two lifeboats: the *Bernard Matthews II* (offshore) and the *Fred Dyble* (inshore). These are run and financed by the Caister Volunteer Lifeboat Service (CVLS), which is supported entirely by public donation.

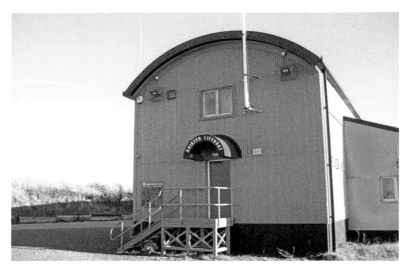

(Paul King)

The RNLI's decision to close the station in 1969 met with a public outcry as a result – perfectly understandable given that, at the time, Caister Lifeboat had saved more lives than any other lifeboat station in Britain. Thus, the lifeboat crew, led by Jack Woodhouse, decided to keep the service running, taking it over themselves the moment that the RNLI withdrew from Caister. The service had two boats to call upon initially, a small fibreglass one to begin with, followed by an inflatable. The CVLS then began fundraising efforts to enable them to purchase a full-size lifeboat for use at the station.

The visitor's centre at the station is open every Wednesday between 10 a.m. and 3 p.m. as well as bank holidays. While visiting, you will also have the opportunity to see (providing it has not been called out, of course) the *Bernard Matthews II*. It is, for the mechanically minded, powered by a twin Volvo Penta engine that gives 450 horsepower to the water jets (propellers are very much a thing of the past if speed is required), lending *Bernard Matthews* a cruising speed of 37 knots. It is currently the fastest offshore lifeboat in the UK.

73 Racecourse, Great Yarmouth
Scene of the infamous soapy tailed pig races.

Great Yarmouth racecourse is situated a few minutes' walk from the beach – handy, therefore, if you have backed some losers and just want to get away and forget about it somewhere. Alternatively, if you've

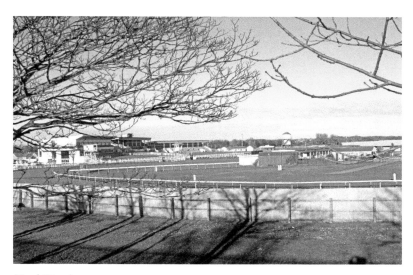

(Paul King)

had enough of all that sun, sand and sea, and want to see if you can win some of your holiday money back – well, you know where to go.

For those of a racing disposition, the course at Great Yarmouth is a left-handed, oblong course of one mile and five furlongs in length. There is an additional straight course, over which races of up to a mile can be run – popular for two-year-olds (horses, not spectators!).

Racing in the town was first granted by the Great Yarmouth Corporation in 1715 (by 1727 there were already 112 racecourses in England; Norfolk's only other racecourse, in Fakenham, didn't open until 1905), when a lease was given for some land to a group of innkeepers in the town who had declared they wanted to stage race meetings there. However, it is possible that racing was already taking place there and that this was a means to make it legitimate, as well as to make them, in the process, some money. In keeping with the town's reputation as a holiday destination that doesn't take itself too seriously, neither did the racing at this early juncture, with the course initially hosting events such as donkey races and the intriguing pursuit of chasing a pig with a soapy tail, as well as its more formal race meetings. A level of decorum was finally established in 1810 when the Racing Calendar began to record meetings that included thoroughbred horses and a sufficient enough purse to interest the respected owners and jockeys of the day. A two-day meeting in the late summer became the highlight of the racecourse's year with, by 1866, a number of regular fixtures at the venue confirming its arrival as an established racecourse in Great Britain.

Like so many other sporting venues in the country, Great Yarmouth racecourse has had to diversify what it has to offer potential customers in order to prosper financially throughout the year. As well as a number of high-end corporate hospitality packages that are available for their more well-heeled clientele, the course also hosts a variety of different events throughout the year, including conferences, concerts, car rallies and antique fairs.

74 River Bure, Great Yarmouth
Tranquil river with a hidden bite.

To take a trip along the route of the River Bure could almost merit a book in itself as it winds its way through many of the most picturesque and prominent locations in Norfolk, much of which are in the Norfolk Broads. It is both the longest and, quite probably, the busiest of all the rivers in the Broads, rising at Aylsham and passing through several towns and villages, including Coltishall, Wroxham, Horning, Stokesby, Runham and, of course, Great Yarmouth, en route to the North Sea at Gorleston. The Bure's two major tributaries are the River Ant (see No. 64) and the River Thurne as well as Muck Fleet, which, as we have already learnt, connects Rollesby, Ormesby and Filby (the Trinity) Broads to the major network.

Navigation of the River Bure starts at Coltishall, where the intrepid sailor can enjoy a peaceful trip all the way downstream

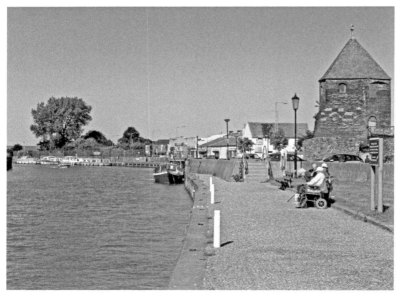

(Mark Oakden at www.tournorfolk.co.uk)

through some glorious Norfolk countryside until it reaches Great Yarmouth, when, just as you might be totally relaxed and at one with the world around you, the navigation becomes a bit tricky! When sailing through Great Yarmouth, if you are cruising between the Rivers Yare, Chet and Waveney and the Bure, Ant and Thurne, then this means that Breydon Water (a large stretch of sheltered estuary on the eastern outskirts of the town) needs to be negotiated. This is a challenge at the best of times but more so if – and this is not an infrequent occurrence on the Norfolk coast – the winds are a little stronger than usual!

It's worth a little deviation from the A149 and the varied delights of Great Yarmouth to look upon the rather earthier sight of the Bure and Breydon Water, which is a reminder that, for all the 'chocolate box' quality of Broadland images, the area can still display an elemental edge. For fans of the novels of Arthur Ransom, the Bure should be something of a point of pilgrimage, for it provides the central location for two of the stories in his famous *Swallows and Amazons* series, namely 'Coot Club' and 'The Big Six'.

75 Britannia Monument, Great Yarmouth
Little known tribute to Norfolk's most famous son.

Yarmouth's Britannia Monument is a commemorative column commissioned and built as a lasting memorial and tribute to Norfolk-born man and naval hero Admiral Horatio Nelson.

The correct name for Nelson's 'other' column is the Norfolk Naval Pillar. It was completed in 1819, the means to erect it having come from funds raised by a committee of local worthies who duly appointed a custodian to 'mind' it once the work was finished. The first person entrusted to the task was James Sharman, a crew member of HMS *Victory*, Nelson's flagship on which he was shot and fatally wounded during the Battle of Trafalgar in 1805. Fittingly, Wilkins was one of the crew members who carried Nelson below Victory's decks after he had been shot.

The monument is in the style of a Doric (a form of classical architecture) column, which is topped by six figures supporting a statue of Britannia who stands atop a globe inscribed with the motto from Nelson's coat of arms, *Palmam Qui Meruit Ferat* ('Let him who has merited it take the palm'). Britannia is, not as one would think, gazing out to sea but, instead, looks inland and towards Burnham Thorpe, the Norfolk village where Nelson was born.

The Norfolk Naval Pillar is 144 feet high (Nelson's Column in London stands at 169 feet) and can be climbed if the visitor is fit and able to ascend the 217 steps to the top. It was, pleasingly, completely restored in time for the 200th anniversary of the Battle of Trafalgar in 2005 and, a year later, removed from English Heritage's Buildings at Risk register.

It seems rather fitting that my modest journey should end at the feet of a man who played such an enormous part in British naval history, and who once, upon embarking on the shore at Great Yarmouth declared, 'I am myself a Norfolk man and glory in being so'.

I think he would have been pleased to know that his county remembers him with pride to this day.

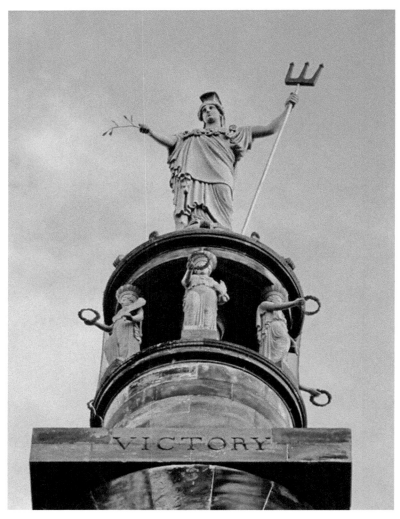

(Adrian S. Pye at www.geograph.org.uk)

Acknowledgements

No book is an island and, in many cases, as with this one, its eventual publication owes as much to all the people who have given their time and selfless generosity as it does to the author himself.

So my thanks and gratitude to all of those who have helped me with *A149 Landmarks* are heartfelt and come with infinite appreciation and gratitude.

First and foremost, thanks to Amberley Publishing and Connor Stait for hearing out my long-time writing indulgence and giving me the commission to make it a reality. This is my fourth book with Amberley; I hope that there are many more to come.

Most of the photographs in this book are courtesy of Nigel Nudds, who, like me, spent his formative years travelling along the A149 for one reason or another. Familiarity never bred contempt in either of us, however; as we took our 'road trip' in late 2016 to put a collection of photographs together, we found ourselves, perhaps inevitably, discovering sights and scenes along the way that we'd never seen before.

I'm also grateful to Mark Oakden of Tour Norfolk for contributing some of the photographs from his outstanding website to this book. You can visit it and contact Mark at www.tournorfolk.co.uk.

I'd like to thank everyone who offered, sent or gave permission for their photographs to be used in this book. I have not, regrettably, been able to use all of them. There is, however, so much more for me to write about Norfolk yet so I am quite sure that I will be in touch with you again about them very soon.

Thanks are also due to the following for their help and advice given in the writing of this book:

Doctor Paul Richards, John Nudds, Councillor Richard Bird, Elaine Bird of *Hunstanton Town and Around* magazine (*www.townandaround. net*), *Iceni Magazine* (www.icenimagazine.co.uk), Professor Fred Cooke of the Castle Rising Heritage Group, Phillip Matthews from Norfolk County Council, Mike Walsh, Pete Goodram, Mark Warner of Jarrold in Cromer, Nick Conrad from BBC Radio Norfolk (foreword), Russell Saunders (proofreading and editing), Janet Lake, and Sarah Povey.

If I have forgotten anyone, rest assured, it is down to forgetfulness and nothing else. Any amendments, corrections or additions required to the entire text will, upon submission to the author, be added to subsequent reprints of this book.

See you on the road!

Edward Couzens-Lake
January 2017
Twitter: @edcouzenslake

About the Author

Born in King's Lynn and raised in a loving family and among wonderful lifelong friends in Brancaster, Edward Couzens-Lake is an author, ghostwriter and broadcaster who currently lives, in very reluctant exile, near Chichester in West Sussex.

He is a lifelong and occasionally long-suffering fan of Norwich City, who has written three books for Amberley Publishing about the Canaries, one of which, *Gossy* (2014) tells the remarkable story of former Norwich player Jeremy Goss. Edward's media work saw him branch out into radio presenting in 2015, and he now regularly produces and presents live shows on two radio stations in the South East.

You can find out more about him and his work at www.couzens-lake.co.uk.

Many of the photographs in this book were taken by Edward's childhood friend Nigel Nudds. Nigel is an extremely talented and versatile photographer whose impressive portfolio of work includes many photographs taken at the live music events that he regularly attends. For more information about Nige please refer to Nigel D Nudds Photography on Facebook.